Dead Hard

Matthew A. Clarke

Planet Bizarro Press

Copyright © [2022] by Matthew A. Clarke

Cover illustration by Sean Clarke

All rights reserved.

Any likeness to persons living or dead is purely coincidental.

No portion of this book may be reproduced in any form without written permission from the publisher or author, except as permitted by U.K. copyright law.

Power doesn't corrupt people. People corrupt power

– *William Gaddis*

Disclaimer

In the age of Tide Pod consumption, I feel a disclaimer is likely necessary.
There are many things in this book that I wouldn't recommend doing in real life. Never ingest something that you're not supposed to in the hopes it will give you special abilities. You will die.
This book contains graphic violence. Lots and lots of graphic violence.

Chapter One

Countless fat, pink bodies rubbed against one another as they fought to get to the front of the pack, to the two humans carrying the fresh meat.

"Damn things are acting as if this isn't the third one this week," Tweak grunted. A plume of dirt kicked up as he dropped the corpse's legs to the floor and wiped his hands down the front of his jeans. "And what's with all this black shit?"

"Beats me. Better hope it ain't toxic to the pigs, though, or the boss will have both our nuts," Guttermouth replied. "Now help me get him over the fence so we can get back to somewhere that doesn't smell like the inside of a hooker's asshole."

Tweak rolled his shoulders, then bent to grab the dead man's ankles.

He jumped away from the body. "The fuck?! His leg just moved!"

"Will you quit screwing around? Marcie was expecting me home an hour ago. I'm gonna be sleeping on the sofa again at this rate."

Tweak edged forward cautiously and bent to peer at the body's glassy eyes. "I'm telling you, his leg flinched when I put my hand on it."

"It's just nerves dying or some shit." Guttermouth waved the dead man's arms in the air between them as if to demonstrate his point. "He's dead. You can't cheat a bullet to the forehead."

The pigs smashed angrily against their metal enclosure, perhaps wondering what was taking so long. One of the runts squealed and kicked out as another bit it on the neck.

Tweak reluctantly wrapped his hands around the corpse's ankles, but as he started to lift, both feet kicked upward and out of his grasp.

"See! I fucking told you!"

"Huh." Guttermouth dropped the man's arms and knelt over his head. He slapped him several times across the cheeks, rolling his head left and right. A thin trickle of black gunk dribbled from the wound on the dead man's forehead, but his eyes remained grey, unfocussed. "Well, I ain't never seen that before, but that don't mean he's still alive."

"Should we put another bullet in him?" Tweak asked, already reaching for the Glock stuffed down the back of his trousers.

"What, and draw unnecessary attention? Even if he *were* alive, he's not gonna be in a minute. Let's get this freak in with the bacon."

Tweak supposed his partner was right. Those pigs were set to tear through anything that ended up on the wrong side of the fence, and they were damned good at it, too. He released his grip on the butt of his pistol and steeled himself to hoist the man's legs for the third time, determined not to flinch no matter what may or may not happen. He grabbed the corpse's ankles roughly, as if that would show it who was boss.

The corpse sat up.

The two gangsters stood in stunned silence as the naked man remained stock-still, his back at ninety degrees to his hips.

"Just a reflex," Guttermouth muttered in such a manner that it was clear even he wasn't buying it. His hands hovered inches from the man's shoulders.

The dead man wrapped his hands around Guttermouth's with the speed of a prize boxer. He pulled down, flinging the screaming gangster over his body and into the other, knocking both to the floor. Before either could fumble their pieces, the dead man was on top of them.

The last thing either of the gangsters saw was the look of pure terror on each other's faces as they rushed toward one another. They smashed together with the power of ten men, two skulls became one.

The man that should be dead stood, peeling them apart like drying glue as he lifted them by the throat in either hand. He tossed the spasming meat to the pigs.

Chapter Two

"What do you mean you can't find those idiots?" Archie 'The Mole' Tanner boomed. He slammed his fist on the walnut desk and knocked a half-inch of ash from the end of his cigar. A thin, yellow-skinned man stepped from the shadows at the back of the office and wordlessly swept the ash into an ashtray before disappearing once more.

The two opposite the desk shifted awkwardly under their boss' imposing stare. Krik eventually spoke up. "We've not been able to get hold of them for days. Pounder got a message from Tweak that evening saying they'd got the guy and were taking him for disposal, but that's the last anyone's heard."

"And I take it you've checked the barn?"

"Boss," Pounder said, rubbing the dog tag around her neck between her forefinger and thumb, "We've had men check the barn, their homes, even the—"

"A 'yes, boss,' would have sufficed."

Pounder noticed a hint of a smirk creeping across Krik's stupid face, poorly applied foundation cracked over the trench-like scars up his cheeks.

As one of the only women in the organization, Sam Pounder felt she had a point to prove and was doing a damned good job of it, too. Ex-army sniper, she had fifty-three red streaks in her cropped grey hair—one for each confirmed kill—and was always looking for an opportunity to add more.

Several beads of sweat fell from Tanner's glistening hairline and slapped audibly on the desk. Instead of blowing his top, as he was prone to do, he set his elbows on the table and buried his face in his hands. The nub of his cigar sizzled as the sweat travelled down his fingers. He tossed it over his shoulder to the man in the shadows. "This was supposed to fix our problems, but someone is clearly intent on pissing me off."

Krik cleared his throat. "We've got men on the streets at all the busiest points in the city. All you have to do is say the word, and we'll have them all over the Trick's turf."

"I WILL DO NO SUCH THING," Tanner said, his red face rising from his palms. A fat, green vein pulsed violently beneath his left eye. "No. I will not risk another turf war so soon after the last. We can't even be sure the guy was working for them, but if he was, I don't want to let them know we're onto them. No. This deal is going through. We'll have some of our men nearby, armed to the teeth, in case they try and pull a fast one. We can worry about the Tricks after."

"Great," Pounder said, a little disappointed. "So, what now?"

Tanner lit a fresh cigar, then sat back in the high-backed chair. "Now, I want you to see who we've got on the east side. I think we need to pay our man Creek a visit. Find out what he knows."

Chapter Three

Jake Creek had been on the force for seventeen years, the first seven of which he'd kept his nose clean, much like the rest of his department.

Until the cuts.

Silver City had fallen on hard times, and the police department was not exempt. In fact, SCPD was one of the first areas to feel it. It started as pay freezes, budget cuts, and zero-hour contracts for new recruits. Eventually, the job made itself so unappealing that only the old and the desperate stuck around (with the exception of one). Then came the layoffs/forced retirements. Further down the line, the entire left side of the rectangular concrete building had exploded in a mysterious 'gas leak', and instead of rebuilding, the force reduced even further.

After Archie Tanner, their finest (and only) undercover agent defected in order to build his own empire from the cash and drugs he'd been stockpiling from busts, the remainder of the force had been tempted to follow suit. Some did, but those that stayed had the deserter's wages split between the rest of them. It wasn't a bad deal, but it still wasn't enough to keep them from playing both sides. Especially when the head of the largest criminal organization in Silver City was an old friend.

"Leave it outside the door," Creek shouted from the bathroom. Why did delivery guys always show at the worst times?

"The Mole sent us," came the reply.

Creek gritted his teeth and wiped faster—he knew better than to keep Tanner's men waiting. What in the hell did those rat bastards want now anyway? He'd only just given them what he knew on the hitman... unless they'd come to reimburse him for his help. Yeah. That had to be it. "Just a minute!" he called as he pulled his trousers up and flushed, rushing out of the bathroom without washing his hands. Two dark outlines stood on the opposite side of the frosted glass door. The kind of ominous shadows you'd cross the street to avoid at night. Creek opened the door.

"Christ almighty, you kill a stray in here?" the first suit said as he barged his way inside. The second followed close behind; his angular face pinched like a bunched napkin.

"Please, do come in," Creek said, shutting the door behind them and gesturing to the front room. "To what do I owe the pleasure?"

"Our boys went missing after taking care of the man you tipped us off to. Boss seems to think you might be able to tell us something about that," said Suit Two. He'd been chewing a wad of gum with his mouth open and, after blowing a small bubble, spat it onto Creek's coffee table.

Asshole. "I told you people everything I know. I thought you might be here to pay me for my services."

"Pay you for what?" Suit One said. "You get paid when the Mole says so. Just like everyone else."

"Well, I don't know what else to tell you, gentlemen. Julio Bandera was brought to the attention of the department several years back and is suspected to be working as a hitman for the Tricks. We had reports of him being spotted several times near the Advance Tomorrow buildings on the city outskirts, potentially casing someone... wait. Did something go wrong?"

"Nah, we got him all right," Suit Two said, popping another stick of gum. "But problem is, we think someone mighta got our guys too."

Creek sneaked a glance over his shoulder, towards the front door. He didn't like the insinuation behind the man's tone. It was dark out, potentially dark enough to slip away, but he'd have to make it to the door first. "I'm sorry to hear that, fellas. Truly. But I'm not sure what any of this has to do with me?"

Suit One strolled to the window overlooking the street outside and began to draw the curtains as he spoke. "We can trust you, right, Creek?"

"Come on now, guys. When have I ever failed to come through?"

"It's funny, is all. You see, you're the only other person who would have known our boys would be making their move on the Trick's guy that night. Of course, I suppose it would be possible that by some off-chance, someone happened to be watching over Mr Bandera."

"Extremely unlikely, but possible," Suit Two chimed in.

"Listen, I don't know what you're getting at, but I don't have anything to gain by withholding anything from Tanner."

Suit One turned to face Creek, a thin beam of light cutting through the crack in the curtains from a nearby streetlight highlighted the dust motes falling across his broad shoulders. Suit One nodded. Suit Two clapped his hands on Creek's shoulders and manoeuvred him into the nearest armchair. Before Creek could voice his protest, Suit Two's fist connected with his jaw with a spray of pinkish phlegm.

"Now you have a better idea of how serious our boss is treating this situation," Suit One said. "Who else did you tell?"

"Please. You're making a big mist—ACH!"

Suit Two shook his fist off, cursing at the cut that had opened along his middle knuckle.

"Careful, Red," Suit One said, "We're only supposed to rough him up a little, not kill the bastard." Red shrugged in indifference. Suit One turned his attention back to Creek. "Sorry about my friend here. He tends to get carried away."

Creek looked up to Suit One with watery eyes. His nose was bent sideways, almost flattened against his left cheek. "I didn't tell anyone else. Why the fuck are you—"

CRACK!

"For christ's sake, Red, hit him somewhere other than the head next time."

"Puhlease," Creek spat, his head lolling forward. A splintered tooth fell between his lips and bounced across the threadbare carpet. It came to rest against the tip of Suit One's polished brogue.

"Is it the law? Someone else in your department finally decided to make a move on the Mole? 'Cos if it ain't the Tricks, that's the only other thing that makes sense to us. Isn't that right, Red?"

"That's right, Sonny."

"Look, if someone else is involved, if your guys were followed, I swear, I had nothing to do with it." Creek drew a whistling breath and held it, anticipating another strike. When it didn't appear to be coming, he added, "I would never do anything to go against Tanner."

Red motioned for Sonny to follow him through to a room across the hall, which happened to be a small, filth-crusted kitchen. "Is he telling the truth?"

"He is, Red. Don't know what he'd have to gain by lying to us, but if he is, he's a better at lying than he is at being a cop. The boss ain't gunna be happy, but he's gotta hear it. Who knows, maybe those other two scored big off the hitman before he copped it and are off blowing the cash in some shithole casino."

"Wouldn't be the first time," Red grumbled in agreement. "Be nice if we could do that one of these days."

"Lose our money on slots?"

"Nah, get away from the illegal stuff for a few days. Relax. Go off the radar."

Sonny looked at Red for a moment as if he was wondering how to respond. Eventually, he settled for shaking his head as he left the room and headed for the front door. "Take care, Officer Creek."

"Be seeing you around," Red called before slamming the door behind them. They pulled the lapels of their coats up to their necks to combat the biting chill of the winter's eve, obscuring the view of their surroundings. In doing so, they lived to see another day.

Chapter Four

Officer Creek's dishevelled residence was nestled between a boarded-up Chinese restaurant and an alleyway littered with hypodermic needles. It was here that the dead man had chosen to wait for the men in the suits to leave the building, studying their attire with keen interest as they reappeared. Neither seemed to notice him as they descended the steps at the front of the property and crossed the street to the parked sedan, but something about them reminded him of the people in the pig barn he'd awoken to. Their suits, and the way they held themselves. These men may be a part of the same organization, which the dead man felt vaguely aware of, but could not quite recall.

After exiting the barn, the dead man found himself in a muddy field and traversed a long, graveled road back toward civilization while attempting to gather his thoughts. He could not remember who he was, or what he was doing with those people that had been about to kill him. He was at a total loss, felt numb both inside and out, and had spent his first night beneath a small bridge with a group of vagrants huddled around a fire barrel. They seemed reluctant to get too near him, but he was unable to formulate any words to ask if they knew him. He'd only killed those men in the barn because they were going to kill him. They thought they'd *already* killed him, if what he thought he'd heard when he was coming to was to be believed, so he'd done it to protect himself. But why would anyone have wanted him dead? Was he a bad person? Bad enough that someone would have ordered a hit on him? As he'd stood there, naked, palms outstretched toward the fiery tongues, his eyes re-

MATTHEW A. CLARKE

focused on a large, blinking billboard across the stagnant river. *Silver City*. The name resonated with something inside of him. That's where he was. But where, exactly, was his home? Did he even own a home?

He'd happened upon Creek's residence by chance, having spent the day wandering the litter-strewn streets searching for some semblance of recognition. People paid him no mind as he stumbled aimlessly, too wrapped up in their conversations or picking apart cigarette butts in the gutters for anything worth salvaging to notice him. He didn't feel the need to put anything in his body. Others were pouring fluids into their mouths or using greasy fingers to push hunks of off-green meats between their teeth, but the dead man could not understand why. As he watched one such person, a young girl with matted hair, a rusted vehicle spun around the corner of the road and almost took her out. The driver stepped out, shouted, waved his arms, before looking around to see if anyone was watching.

The dead man stepped back into the alley's shadows, kicking aside a syringe with his bare foot.

The man kicked the girl in the shin before leaving her rolling in the middle of the road.

Officer Creek.

The name ran across his mind like a digital display in a shop window. He knew the name but couldn't put his finger on the 'why'. The thought that he might be a bad man once again crossed his mind—had the officer arrested him at some point? No, that wasn't it.

Time didn't mean anything to him, so the dead man remained in the alley, watching the property until a black blanket was drawn across the sky and the first of many stars twinkled into existence. It was only when the car pulled up across the road, and the thuggish-looking strangers entered the property that something clicked.

Officer Creek had been watching him. He hadn't thought anything of it at the time, but he'd noticed the man several times in the days leading up to his attack. Whenever they'd made eye contact, Creek had either walked away, pretending to get a phone call, or disappeared into a nearby building. It had to mean something. It was the best he had to go on.

The dead man mounted the stairs to Officer Creek's front door. He could hear laughter from inside the house, which was then accompanied by a short bass

DEAD HARD

lick. *A television.* Somewhere beneath it all was a man's voice, strained. Angry. Yet he was fairly sure Creek was alone in the property. The dead man knocked twice, and the house fell silent. A hunched figure approached on the other side of the cracked glass.

"Go away."

The dead man did not go away.

"Are you deaf? I said fuck off."

With the dead man standing his ground, Officer Creek opened the door just wide enough to show his ruined face and the top of a baseball bat in the opening.

The dead man shouldered the door, and it snapped horizontally in a hail of glass and wood. The weapon was out of the officer's hands before he'd even hit the floor. "Y-you," he stuttered, scampering backwards on his behind. A faint, deep voice that had been shouting something down the other end of Creek's phone was muted as the dead man crushed it underfoot. "You're supposed to be dead." The policeman found his footing and retreated into a room across the cramped hall before the dead man could close the gap, slamming the door behind him.

The dead man tore through the flimsy wood with as much effort as snapping a toothpick but was immediately knocked sideways by a loud blast. There was a dull pressure just above his abdomen. No pain. He growled, took another step forward, and another three bullets punctured his shoulder, pectoral, and upper thigh. Creek screamed as the dead man wrapped his hands around his own—all while shrugging off yet another bullet—and crushed both his fingers, and the weapon, into an indecipherable mess.

"Please," he managed to gasp while hot white pain washed over him, "I'll tell you anything!"

The dead man pushed Creek to the floor and stood over him, chest heaving, while he fingered the puncture wound on his chest. A thick, black paste had already begun to solidify around the hole in his puckered skin. It had no smell, no taste. He looked to the bloodied officer, wondering what he should do now. Without communication, he could do nothing but kill, so he attempted to form words while Creek watched on, eyes pinched.

"They made me tell them, I swear to God. The Mole would have had me killed if I didn't tell them what I knew about you."

MATTHEW A. CLARKE

"Who me?" the dead man said.

"You . . . you don't know who you are?" The officer continued once it became apparent he would not get a response: "Your name is Bandera. Julio Bandera. You're a shooter for the Tricks."

The name meant nothing to the dead man. He was reasonably sure he was not who he was being told he was. He shook his head. "No."

"No? Wait. You're not. Are you? You look a hell of a lot like him," Creek propped himself up on one elbow, wiped the blood from his eyes. "But Bandera has a scar on his upper lip. What the fuck are you?"

But the dead man wasn't listening. A memory hit him like a child on the main road in a Stephen King novel.

His name was Benjamin. Benjamin Hard. He lived north of the city, alone, in a small, ground floor apartment. Further memories were flooding him from all directions, making it near impossible to focus on Officer Creek. He had to get out of there. Find somewhere safe he could sit in silence and allow the mental tapestry to unravel undisturbed. He shook himself off and kicked Creek's elbow out from under him.

"NO!"

Hard brought his knee up to his waist, then brought his heel down on the middle of the policeman's face.

Chapter Five

Benjamin Hard had first suspected he was being trailed when he popped into Dell's Diner to grab a coffee to go and noticed the same black transit down the street as the one that had been outside his house that morning. Call it paranoia if you will, but in Silver City, it pays to be paranoid.

Hard was one of the more fortunate residents. However, he made a point of never making it too obvious for fear of attracting unsavoury characters. He prided himself on his awareness of his surroundings and his ability to read a situation. Unlike many of the residents of Silver City—named after the mines that had long since closed down—he'd been fortunate to have received a decent education and was able to drop right into a job as a lab technician at Advance Tomorrow once the silver trade dried up. The pay was a little above average, and his experience working with precious metals meant he was a better candidate for the position than many others. The majority were not so fortunate—with the sudden rise in unemployment and severe lack of jobs to meet it, debt and crime became commonplace.

Benjamin saw Creek on several occasions the following week, and although his unease was growing, he couldn't see what a man of the law (no matter how crooked) would want with him. After all, he was a fairly boring guy, living a fairly boring life. Had he trusted his gut a little more and taken a break from the city, he may have avoided what was coming his way.

MATTHEW A. CLARKE

"You sure you're okay to lock up, Benny?" Lang asked with a cheeky smile. He hated it when she called him that, but he was powerless to do anything but go along with it.

Khoai Lang had been working with Benjamin for a little over five years, and their relationship had evolved the way one may when two singles spend prolonged periods alone together. The first time he'd seen her, he remembered thinking she was a little weird and definitely not his type, which was good because it meant he'd be able to focus on his work. But as time went on, and he grew used to the way she would insist on brushing her teeth between every meal or drink, or the way she liked to nibble on chalk sticks (he eventually grew to adore these things), he realized he was starting to fall for her. He'd eventually plucked up the courage to ask her to accompany him for a drink after work. They'd been engaged for three months.

Benjamin assured her he'd be fine and saw her off with a quick peck on the lips and a playful slap on the bum. There wasn't much left to do; shut down the computers, check the supplies were secure, and perform a quick walk around the wing to ensure everyone was out. He was on his knees in the loading bay, checking the locks on the shutters, when he heard the tell-tale sound of a gun's hammer being cocked not two feet from the back of his head.

"Undo the lock. Ah, ah. No peeking," said a male voice as Hard went to turn around. "Just undo the padlock."

Hard had no choice but to oblige. He slowly reached for the ring of keys on his belt, released the padlock, and placed it on the floor next to him.

"Good. Now open the shutters."

"Take what you want. Just let me—"

"Open the shutters if you want to be able to walk out of here."

Hard adopted a squatting position and lifted the metal shutters to head height. A vehicle was idling outside, its headlight facing inward, blinding him. The driver's door opened, and a second figure stepped out.

"Outside. Now."

Benjamin was reluctant to go anywhere with these people. In fact, an overwhelming feeling in the pit of his stomach told him that he was as good as dead the moment he stepped outside. Without thinking, he spun and broke into a sprint, running for the door, but froze as a bullet whistled past his left ear and buried itself in one of the large plastic drums next to him. Graphene powder erupted from the small hole like a rancid waterfall, spilling across his work boots.

"DON'T FUCKING MOVE."

Hard threw himself at the tall stack of containers, knocking the bottom one sidewards and triggering a domino effect on the others as another gunshot rang out. Within seconds he was buried, trying to catch his breath as black powder cascaded all around him, filling his mouth, ears, and nose. Muted shouting, moving closer. What was he thinking? That the plastic drums would have knocked the man over and caused him to lose his gun? That shit only worked in cartoons and movies. All he'd succeeded in doing was trapping himself, reducing his chances of a relatively painless (he hoped) death, and almost guaranteeing a drawn-out death by suffocation.

At least, that's what he thought.

A moment later, the barrel that had landed across his face was lifted away. A tsunami of graphene powder flooded from its ruptured lid. Hard inhaled a lungful and immediately began choking.

"I got him, Guttermouth," said the man training a gun on Benjamin's head.

"Don't use my name, you fuckin' idiot."

"Why not? There's no one else here."

The man named Guttermouth rubbed his forehead with the palm of his hand. "Whatever. Just do it."

"You want to grab a beer after thi—"

There was a flash of white light.

Chapter Six

Felix Schafter had seen some grizzly sights in the two years he'd been working for SCPD, but this was something else. Hell, even the skinned politician they'd found in the bottom of a grit bin last winter hadn't been enough to evoke a reaction. It had caused his partner to double over and vomit across the cracked pavement, and the woman who'd discovered it to later blow her brains out mid-shift at Dell's Diner.

His partner wouldn't be vomiting anymore, though. He wouldn't be doing *anything* anymore.

Jake Creek's face was caved inward. Blunt force trauma. A club hammer, or a crowbar, if he'd had to bet. If you'd told him a single downward heel strike had caused the devastation, he wouldn't have believed it.

Schafter placed the back of his hand against his mouth as he moved closer to the body. He'd been sharing a squad car with the guy not four hours ago, and it was hard to believe this was even the same person. Creek's nose had essentially inverted, disappearing into the pond of blood and pus that had pooled in the crater of his face. His upper lip had been pulled upward into a vicious sneer, and the flesh around his eyes had been dragged down, with his lower eyelids under the blood line, collecting it.

"Maybe you should take the rest of the night off," Lime said, in an unusual display of emotion.

Schafter glared at the man as several figures in white coveralls passed through the ruined front door—the door itself was propped up against the wall in the hall—and moved about the area, taking photographs of bloodstains and the body. Schafter knew they should get out of their way but, at the same time, was reluctant. If he didn't stick around to keep an eye on things, he knew there was a good chance something would be 'accidentally' missed or contaminated. It was no secret that his partner had been involved with the Mole and his gang of cronies. Hell, most of the force were, which may explain Lime's general sense of unease at the situation. Schafter was perhaps the only police officer that wasn't willing to accept bribes or help things disappear from the evidence room (if it even made it that far in the first place). Sure, there were crimes that *had* to be solved to keep up appearances, such as the occasional theft perpetrated by a desperate vagrant or the evasion of city taxes, but Archie 'The Mole' Tanner had them all in his pocket for the most part.

Schafter had always wanted to be a policeman growing up. When he was four, his father had joined the force, only to resign a few years later for reasons he was unwilling to disclose to his son. (Schafter's mother would later tell him his father simply couldn't handle watching the city consistently fail its people). Luke Schafter had been murdered soon after. A month later, Penelope Schafter died of a broken heart (or so Felix believed), leaving Felix to fend for himself. It wasn't easy, but he was better off than most with his parent's property already paid off. By the age of twenty-three, he was on the force. It had been surprisingly easy—with poor pay for the hours and further cuts on the horizon, they'd practically dragged him in the front door.

Schafter had been partnered with the foul-tempered Creek straight out of basic training and had been suffering under him since. They hated each other and made no attempt to disguise it. No matter how bad things got, Schafter refused to take the carrots dangled in front of his face in exchange for looking the other way and would do his best to make sure Creek was not doing anything he shouldn't be either. Clearly, the long conversations he'd had with the man while cruising the city had fallen on deaf ears, for he had been murdered by the very people he'd been willing to ruin his reputation for.

"Have the neighbour who reported the gunshots brought in for questioning," Schafter said to Lime without taking his eyes off his dead partner. "We need to know if they saw anyone."

DEAD HARD

"You don't tell me what to do," Lime said flatly, leaving the room.

The response didn't surprise him. Even when it was one of their own, they were hesitant to investigate properly, to make waves. Everyone was only concerned with watching their own back. Sometimes it felt as if he were facing overwhelming odds, and he would wonder what the point of even trying was. But then he remembered why he joined the SCPD in the first place; to fix the city his parents and their parents before them had grown up in, or die trying.

He scoured the scene the best he could without touching anything. There was a lot of blood. Not only in this room, but up the hall to the front door, too. It was possible that it belonged to the assailants, but he wouldn't know until (*if*) the results came in. The neighbour had mentioned hearing several shots fired, and the call handler had even heard one herself while on the phone with them. So, where were the bullets? None appeared to have struck the walls or surrounding furniture, which could only mean they'd hit their intended target/s. He'd have to call each of the hospitals around the city and check for people admitted with bullet wounds. (Although he doubted that would get him very far—not only were gunshot wounds fairly routine, but those that received them from doing things they perhaps shouldn't have been would choose to deal with it themselves, or visit one of many illegal underground medics that were known to operate in the city).

Another thing of concern: Creek's hands and what had presumably been his service weapon. Crushed with such force that the bones were all but gone, and the fatty flesh of his fingers had melded with the flattened butt of the gun. He'd never seen anything like it. Then there were the black stains. Small smears on the carpet and more on the wall nearest the door. He supposed gunpowder residue was a possibility, but again, wouldn't know until later. "Make sure to get a swab of that," he said to the nearest white-suited figure, pointing to one of the larger black smears on the floor. The figure nodded, and Schafter hung around until he'd seen them swab and bag a sample of the substance. He also made a note of the man's face. Just in case.

Schafter stood on his partner's front step and lit a cigarette, gazing through the rising smoke at the piss-stained street before him. A homeless drunk groaned and rolled over on his sheet of cardboard. Had things always been like this? He had great memories of his childhood on these very streets, playing with his parents and riding bikes, but had he been too young to notice how it really was?

As he started down the steps, he noticed a distraught-looking woman peering from the edge of the curtain of the house across the road. Likely the neighbour who'd called it in. No surprise that Lime had left without bothering to bring her in—just one more thing he'd have to do himself.

"Go home, son."

Detective Schafter cast a half-glance to Chief Feltstone, then back to the frail woman on the other side of the one-way mirror. "I don't need to go home. This is my case. I'm seeing it through."

"You're too close to it. We can't afford to have any slip-ups when dealing with the death of one of our own."

Like my father? Schafter wanted to say. He knew the real reason the Chief wanted him out of the way was so he could get the entire mess swept under the rug. Instead, he replied: "I'm fine, sir. I owe it to Officer Creek to bring his killer to justice."

The Chief moved between the detective and the window and folded his thick arms across his puffed chest. "I wasn't asking. You're off the case. You've been signed off, paid, for the next few days. Shock can have a funny way of rearing its ugly head when you least suspect it, and in this line of work, that's something we can't afford."

Schafter knew to protest further would be futile. He noticed the rest of the men in the room were eyeballing him too. Not a decent one among them, each daring him to slip up and do something that would get him fired, or worse.

He wouldn't give them the satisfaction.

Feltstone unfolded his arms and placed them on his hips. "Come back next week. I should have a new hire for you to partner with by then. My advice? Use this time to get some rest. Stay out of trouble. You've been working far too hard recently."

"Sir."

A snicker came from somewhere behind him. Schafter turned to see who it was, but most of those present were smirking. Yet another battle he could not win.

It was important to choose them wisely.

Chapter Seven

The Mole set the earpiece of his rotary phone gently on the desk and took a deep, measured breath. "FUCK! FUCKIN' NO GOOD PIECES OF SHIT!"

He picked the receiver up again. Palms sweaty. Knuckles white.

"My condolences. I'll have a word with them. If it turns out they're responsible, you can be assured they will be dealt with accordingly. Yes. Okay. Goodbye."

"You wish for me to bring them in, boss?" said the skinny man in the shadows.

The Mole waved a chubby hand at the door dismissively, as if to say he really didn't want the pair of knuckleheads in his sight but at the same time understood this was a situation that needed to be dealt with swiftly. The skinny man crossed the room and beckoned the two men waiting in the corridor inside.

Red and Sonny looked at one another like a pair of school kids outside the principal's office, each waiting for the other to enter first. Eventually, Sonny took the lead, and the pair seated themselves in the faded brown leather wing chairs opposite Tanner. "Is there a problem, Boss?" Sonny asked meekly.

"Let me think," Tanner began. His fingers, adorned with enough fat, golden rings to make an ancient Egyptian ruler envious, thrummed a steady tattoo on the scarred wood between them. "Did I, or did I not, ask you explicitly *not* to do anything to our dearly departed Officer Creek that could have brought tension between our organization and the SCPD?"

"Dearly departed?"

"We didn't—" Red began.

Tanner stopped drumming abruptly. "I sent you there to get information. NOT TO KILL THE PARTNER OF THE ONLY STRAIGHT FUCKING COP IN TOWN!"

Sonny forced himself to look his boss directly in his deep-set eyes. "The man was alive when we left him."

"Alive? Tell me, how long did you expect him to be alive after what you did to him?"

"I think there's been a misunderstanding," Red said. "We roughed him up a little, but nothing that would have caused lasting damage."

"A misunderstanding?" Tanner propped his elbows on the desk and steepled his fingers. "Yes. A simple misunderstanding. A simple misunderstanding that could have ruined our arrangement with the SCPD irreversibly and force my hand in dealing with Schafter once and for all." Tanner raised the index finger on his left hand, barely perceptibly.

Red gasped as the skinny man drove a curved blade through his skull and into his brain.

Sonny yelped, kicked away from the desk, knocking his chair to the floor as his partner slid down his seat, eyes rolling, spasming and mouthing something wordlessly. "Tanner, listen," Sonny pleaded, his eyes frantically scanning the darkness that ran the circumference of the dim office in search of the skinny man. "You have to believe me. Creek took a few punches, nothing more. He was still standing when we left."

Red uttered a wet gurgle as he slid off the chair and face planted the floor.

Tanner motioned for Sonny to sit. Sonny did so reluctantly, lifting his right foot to avoid the spreading pool of crimson flowing freely from Red's nostrils.

"I always liked Red," Tanner said with nonchalance, "I like you a whole lot less."

The man was lying, but Sonny knew better than to push his luck. "Boss. Someone must have been there after us. They're playing us. Have to be."

"The Tricks," the Mole muttered. He fished a cigar from the wooden box atop his desk and lit it with a gold-plated zippo. "Did Officer Creek at least tell you anything useful?"

"He denied all knowledge. He had no idea what happened to our guys."

"And you're sure he was telling the truth?"

Sonny nodded.

Tanner grunted. "Make some calls. We're stepping up security."

For a moment, Sonny thought his boss was addressing him—he had no idea who to call but was too nervous to say so. But then the door shut behind him, and he realized the skinny man had left the room.

The Mole tilted his head back and blew a thick plume of cigar smoke across the tar-stained ceiling before continuing. "If our hands are clean on this, that only leaves the Tricks, or the SCPD. It makes no sense."

Sonny was content to remain silent for the rest of the meeting, but after a moment, it became clear that the Mole was expecting a response. "Could be the Tricks are looking to make a move on our operation . . . if they could damage our relationship with the SCPD . . . I don't know, boss."

The Mole lifted the corner of his lips as he tapped the ash from the end of his cigar onto the desk. The skinny man appeared from the shadows behind him and swept it into a glass ashtray with a small brush. Sonny hadn't even heard him re-enter the room.

"You might be right. I'm going to arrange a meeting with Feltstone. Face to face. I want you there. We're going to see if we can't hash this out and figure out a way to take the Tricks out of the equation once and for all."

"Fuckin' asshole," Sonny muttered as he ran his tie under the hot tap. The water eventually turned from off-pink to clear as the last of his friend's blood disappeared down the plughole. He moved to the hand drier, holding it beneath the lukewarm air as he studied himself in the mirror to the left. Dark circles enveloped his bloodshot eyes. He couldn't remember the last time he'd got more than five hours of sleep, not that it was any grounds for complaint—the boss had been riding the entire organization hard in the weeks prior. Sonny knew the Mole got to where he was now with a little hard work and a lot of backstabbing, but it could be argued that it came at the cost of his sanity.

MATTHEW A. CLARKE

He'd watched as the man he once admired became increasingly paranoid and distrustful of all those around him. He'd known it was only a matter of time before he started whacking members of his own crew, he just didn't expect it to be so soon, nor someone he'd spent so much time with. Personally, Sonny didn't think the Tricks were involved in whatever was going on at all but knew better than to tell Tanner he was on the wrong tracks. The Tricks wouldn't be stupid enough to make a move on their turf. Would they? It was more likely that it was one of their own, working to dismantle their organization from the inside and strike when the time was right—the same way Tanner had done when he took over. But there was no chance in hell he would mention that to the man, especially after the shit show he'd just witnessed.

And don't even get him started on that freak, the skinny man.

Chapter Eight

It was raining hard enough to drown out the ambient noises of the neighbourhood but not enough for Benjamin to feel it. Truth be told, he was coming to realise he couldn't feel much of *anything* anymore. Temperature, pain, hunger. The only sensation he had, the only thing that convinced him he wasn't entirely dead, was the ability to feel pressure. His wounds had crusted over, the black gunky fluid crystalized to form a gauze to pull the skin around the holes closer together, but he was sure the bullets were still inside him. How any of this was possible, he did not know, but the less he thought about it, the more he was able to function with a semblance of normality.

After transforming Officer Creek's face into a strawberry punch bowl, Hard had performed a quick sweep of his place for anything of use. He'd found a spare SCPD uniform in a broken wardrobe next to a single bed and had been using it to move about the city without attracting too much attention. People were just as afraid of the police as they were of the Mole and his crew and would do their best to keep their distance wherever possible. He'd also taken Creek's mobile phone. It was locked and required a password to use. Had he been entirely in control of his thought process, he may have left it behind for fear of being tracked, but it turned out to be a good thing he hadn't, as the phone vibrated against his chest. The locked screen lit up.

Tino's fishery store. 10 PM tomorrow.

It was a group text from Randal Feltstone, the Chief of the SCPD. Another notorious scumbag that wouldn't be missed by his own family.

MATTHEW A. CLARKE

The phone bleeped twice, and a battery icon appeared on the top-right of the screen before the thing shut off entirely. Benjamin cast it aside like an unwanted child, watched as a small river of rainfall carried it down the nearest drain, and turned his attention back to the woman in the window.

He hadn't known why he'd felt compelled to visit this side of the city until he got here, only that it was familiar to him and might go some way to working out why what had happened to him, had.

Lang was visibly distressed as she moved aimlessly from room to room. Although it was dark out, she'd left the curtains open and Benjamin—as well as anyone else that happened to pass—was able to see straight inside. She spent several minutes folding clothes in the front room, then shook them all out and folded them again. Then she moved to the kitchen, and it was here that he could see tears rolling down her cheeks, although her face, for the most part, appeared placid.

Footsteps approached from his right, splashing as they kicked through puddles. Benjamin took a step back to allow the man with the little girl to pass. The man lifted his umbrella so as not to collide with him, then looked to the lit-up house, back to Benjamin, unable to disguise his disapproval. Benjamin paid him no mind. His eyes remained on Lang as she poured herself a mug of black coffee. Instead of drinking any, she simply brought it up to her chin with both hands and inhaled the curling steam that rose from it with her eyes closed. He wanted more than anything to go to her, set her mind at ease, and climb into a soft bed together, but knew they couldn't.

Something was wrong with him. There was little doubt about that. Whatever had happened between the failed kidnapping and his regaining consciousness in the pig barn was to blame, but that blank period would remain that way. Unless . . . surely not? The last thing he could remember before he was looking down the barrel of that fucker's revolver was drowning. Or was choking a better word for it? He'd been suffocating on the very thing he'd been working with every day. Graphene. The wonder metal. It was the cancer cure of the construction world. High elasticity and flexibility. Incredible resistance and hardness in even the thinnest layer. But it was ridiculous to think that ingestion of a significant amount could re-

DEAD HARD

sult in . . . whatever the hell this was. Wasn't it? Benjamin was certain it had never actually been done before, so there would be nothing to back that (reasonable) assumption up. Ingestion of a toxic chemical such as that would almost one hundred percent more likely end up in a slow, agonizing death.

Lang left the kitchen and disappeared for a few moments, only to reappear upstairs, in the master bedroom. It wasn't much to look at from the outside—pale yellow walls, a single wardrobe and a bare bulb—but once you were in there, it had an atmosphere you simply couldn't anticipate. He remembered the first time he'd been invited inside, up to her room. Lang used a combination of traditional Vietnamese music, scented candles and small paintings to create her temple of serenity. Benjamin thought he could hear the soft call of a string instrument through the pounding of the rain, only today, the music seemed to be doing little to calm her. Lang slipped her blouse over her head and used it to blot her eyes. Her hands then disappeared behind her back, to the clasp of her bra, and only then did it seem to register that she was on show—a dangerous thing indeed for a woman living alone in this city.

Lang grabbed a fistful of the curtains on either side of her window and froze, looking directly at him. Surely, there was no possible way she could see him, with the streetlights on this side of the road busted and the rain coming down as heavy as it was. Benjamin held his ground, looking up at her, wishing he could show himself as she moved closer to the glass. So close that her nose was practically squished against it, and her breath was blotting patches of fog. Still, he remained. In a way, he hoped she would see him if only to tell him everything was going to be okay, but he knew he couldn't allow it. He had things to do. Things that needed to be dealt with on his own. There was a good chance there were still dangerous people after him, and he'd never be able to forgive himself if anything happened to her.

MATTHEW A. CLARKE

Lang paced aimlessly, putting off simple tasks such as eating or showering so she could be ready to answer the phone should it suddenly ring, or the door should Benny suddenly knock. It was completely out of character for him to be off work, especially without letting her know how he was, but she'd been calling and texting constantly for the last couple of days to no avail. Eventually, she'd convinced herself that he must be dealing with an emergency and worried she'd be annoying him.

Her boss hadn't been worried. Annoyed, sure, but not worried. There was nothing to suggest he hadn't gone straight home after locking up that night, aside from a few spilt barrels of graphene powder. Even Lang had to agree that was likely caused by a vagrant, of which plenty were hanging about the estate (and everywhere else for that matter). It wasn't the first time, and it certainly wouldn't be the last. No matter how well they checked the place before leaving, those bastards were like rats; always managed to find their way in.

She retreated to her bedroom and selected a relaxing song from her phone to play through the wireless speaker on her nightstand.

She decided that she would go to the police in the morning and immediately felt a little better. She was aware as any that the SCPD did not have a great track record for finding missing people, but perhaps with a little financial persuasion . . . yes. She would gather any money she had and offer it in return for their services.

Lang slipped out of her cotton blouse and dropped it into the wash basket. It was only as she was about to release her bra that she snapped from her fugue state and realized she was on show to the world. She cringed at the thought of someone standing across the street, touching themselves—something that had been at the back of her mind since some local kids busted the streetlight. Only this time, as she moved to cover herself, there *was* someone there.

She flinched a little, squinting through the droplets running down her window and expecting the figure to run away, but they didn't. Was it a police officer? It certainly looked like the uniform, and she could make out the golden badge on his chest as it caught the glare of the moon. It *was* a police officer. Had her boss already called the police? Were they watching her movements, investigating Benny's disappearance without her intervention?

Benny.

There was no reason why Benjamin would be standing across the street in the pouring rain wearing a regulation police uniform, but she'd learned to trust her eyes and her instincts from her years of working with dangerous chemicals. "Benny?" she said, feeling ridiculous for even suggesting it could be him. The rain must be distorting her vision more than she thought. That must be it. If it *were* him, why wouldn't he just come to the door?

Lang decided there was only one way she would ever know for sure. She snatched her blouse back out of the wash basket and threw it on while crossing the landing to the stairs. She descended, flung the front door open, and stepped out into the rain. Her hair was immediately plastered to her face. She used both hands to sweep it behind her ears as she crossed the road to the location she'd seen the man. Other than a mobile phone, there was nothing to suggest that anyone had ever been there. Lang pocketed the device and hurried back inside to safety.

Chapter Nine

Felix Schafter picked up a case of beer on the way home and had just put the first to his lips when his phone started vibrating. He considered ignoring it—with no real friends or family, it was likely to be a spammer—but checked the caller ID out of curiosity.

Number withheld.

He hung up, placed the phone on the worktop, and took a long draw of his beverage. The phone went off again, dancing a slow pirouette on the faux marble laminate. This time he answered, on the off chance it was the Chief, telling him he'd changed his mind.

"Detective Schafter?" asked a squeaky voice on the other end of the line.

Schafter could hear the clinking of glass and people moving about in the background.

"Who's this?"

"Tim," the man paused a second, waiting for recognition. "Tim Philips. Forensics."

Schafter recalled the name but couldn't match a face to it. "What do you want, Philips?" he asked, making little effort to disguise his impatience.

"Yes. Uh. I've got the results of the black substance we recovered from the crime scene today."

"That was quick," he said, humouring the guy. "Well?"

"It's graphene powder. Not something you'd expect to see at a crime scene, especially mixed with blood."

Schafter took another sip of beer, wondering how that information could be of any use to him. He thanked Philips, hung up, and retrieved his laptop from the front room. After a quick internet search, he was no better off—there was nothing he could see that would explain why the substance would be at the scene at all. He hit the search bar once more, this time typing 'Graphene Silver City' and was surprised to find the top hit was relevant.

It was an old article from several years ago posted by an obscure news site. Only a few paragraphs long, no images, stating that Advance Tomorrow had bought Silver City's industrial area at a steal after the silver mines had closed and businesses moved elsewhere in search of the next profitable venture. The article went on to say that Advance Tomorrow dealt primarily with graphene research and application, which can be highly profitable, but also that it was unlikely to mean much for the citizens of the dying city. They would not benefit directly from any successes the company achieved. The article then said other tech and pharmaceutical companies were looking to invest in the area because land was cheap (for an outsider).

Schafter shut the laptop and sat there for a moment with his hands folded atop it.

What could Advance Tomorrow possibly have to do with the murder of a police officer? And not only that—but to be so careless as to leave graphene all over the crime scene? None of it made any sense, and none of it *ever* would, if he left it up to the officers that had been assigned to the case. Hell, he'd be surprised if the paperwork hadn't already 'slipped' into the shredder, if it had even been done in the first place. He didn't owe his partner anything. He'd never liked the guy (and was secretly a little pleased that there was one less person such as Creek on the force, if he were being honest with himself), but that was beside the point. A murder is a murder, no matter the specifics, and it had to be dealt with appropriately.

With nothing but time on his hands, Felix decided he would take a trip up to the industrial estate in the morning and have a little poke around. He doubted he'd bump into anyone from the station there, even though Philips must have passed the lab results on to the rest of them by now. People were too scared of one another to act. He was sick of the Mole having his way with the city. It was time to act. If not for Creek, then for his own father. If he had to drive a stake through the black heart of corruption on his own, then so be it. Enough was enough. Nothing would ever change if the change wasn't forced. He no longer cared if it cost him his job.

DEAD HARD

Or his life.

Chapter Ten

The blacked-out BMW slowed once the warehouse came into sight. The atmosphere inside the vehicle was charged, everyone on edge, whether they'd admit it or not. The rest of the three-car convoy pulled up behind them, and Tanner and his men gathered on the concrete expanse outside.

A couple of unmarked vehicles sat in the parking bays outside the drab building, along with a single SCPD sedan. *Idiots*, Tanner thought. Why in the fuck's name would they think it was a good idea to bring that to a meet? As he adjusted his silver cufflinks, he said, "Keep sharp." Then, to Krik, "And what the fuck have you done to your face? You ain't foolin' no one, pretty boy."

Krik subconsciously touched a hand to the old wound on the left side of his face. He knew his attempt at covering his scars wasn't perfect, but he'd hoped to at least disguise them enough to be passable. The Chelsea smile had affected his confidence more than the actual act itself. He'd tried plastic surgery, but that had only made it worse. Fake skin had been laughably bad, and makeup was his latest approach. He considered wiping the blusher away with the back of his suit sleeve but ultimately decided against it—just because his boss saw straight through it didn't mean someone that didn't know him all too well would.

Pounder detached the walkie from her belt and brought it to her mouth. "We good?" she whispered.

"All clear," came the response.

Satisfied every possible entry point was covered, Tanner directed his crew inside.

MATTHEW A. CLARKE

The warehouse was spacious, but that was about the only thing going for it. The walls were damp, and the bulbs were dim, strangled by cobwebs and age. Countless stacks of wooden crates lined the walls and created several aisles throughout the building, many of which reeked of marine life. Each was stamped with a large, black octopus with a cigar clutched in each tentacle. Tanner hated meeting in this shithole, but it was the best place for it should anything go wrong—plenty of ice and containers around, and the owner was in on the deal. All it had taken was a little cash and heroin, and their organization had the perfect front to store and distribute their product (be that drugs, or weaponry, whatever was cheapest to get hold of at the time).

"Archie," Feltstone said as he looked up from his tablet. Several men flanked him on either side. Even in civilian clothing, Tanner could tell they were police from the sucked-in beer guts and the fear in their eyes.

Good, he thought, they *should* be nervous. Someone was trying to diddle him bad, and they'd better hope it ain't them. "Randal," he responded, eyeballing a service pistol's outline beneath the man's bowler shirt. "I thought we agreed to meet unarmed. You know, in the interest of . . . everyone's safety."

"Forgive me, old friend," the Chief said, straightening his shirt over the service pistol, "one can never be too careful. Old habits and all. I can assure you the rest of my men are unarmed."

Tanner would have been tempted to call the entire thing off right then, if he didn't already have the building surrounded by enough firepower to bring down a tank. He could already tell that the man was as distrustful of him as he was of them. He decided to let it slide. "I've requested this meeting to discuss our arrangement going forward and where it leaves us with what's been going on in our beautiful city."

Feltstone grunted. "We're working around the clock to locate your missing men. You'll be the first to know when we find something. But while we're on the subject of suspicious activities, do you have any further information on Creek? Now, you know I wouldn't accuse you of anything, so don't take this the wrong way, but we both know he was skimming a little off the side."

The Mole felt heat rising in his cheeks but did his best to control himself. Had he known the officer was profiting personally off their arrangement, he would have had him whacked long ago. A foreign emotion bubbled in his stomach as he thought of how he'd had Red killed. Remorse? A moment later, it was gone. He cleared his throat. "I think whoever is responsible for the death of your man is the same as whoever's responsible for mine. People don't go missing in our line of work unless they're undercovers, or dead, and we both know it ain't gonna be the first option."

Feltstone's left eye twitched, his interest piqued. "The Tricks? Is that what you're insinuating? They left town months ago. Shortly after you made an example of the East Siders."

"Be that as it may, your man Creek informed us not long before his demise that he'd spotted one of their hitmen, one Julio Bandera, in several locations around the city. I had two of my guys take him out—just to be safe, you see—which were coincidentally those very same guys who disappeared shortly afterwards."

The warehouse fell silent as Tanner let the information sink in. Somewhere beyond one of the cracked windowpanes, a stray cat wailed. Krik flinched.

Feltstone lowered his voice as if the walls were closing in on them, "You think it was a setup."

Tanner spread his palms.

"If the Tricks are indeed planning on making a move on Silver City, this is exactly the kind of thing I'd expect them to do."

Relaxing a little now, Tanner lit a cigar as he waited for him to continue.

"If they take out guys on both sides of the equation, rather than say, just yours, they could make us work against one another rather than as a single unit, then swoop in while we're distracted and clear up."

"All shipments must be postponed until they've been dealt with adequately," Tanner said, standing taller than his stunted stature would allow.

Several mumblings of disagreement floated about the warehouse.

"Boss," Sonny whispered, eyeballing the nearest officer. He considered bringing up the theory that it was someone else altogether, but ultimately decided against it. He didn't want that kind of heat on him at an already tense moment.

Tanner remained stoic. They stood to lose just as much as the SCPD if the business at the docks were to pause, even momentarily. They'd been running a smooth operation for years, never straying from the modus operandi. That was the deal. Their suppliers didn't care what was going on at their end. They would take the disruption personally, and it would cause chaos all the way down the chain.

Chief Feltstone remained impassive, unreadable. After a moment, he turned to address his men. "Tanner is right. It's too risky to continue until we've dealt with the problem at its core." To Tanner: "I'll send a team of our best men to deal with them. It's better we catch them off guard on their own turf than waiting for the rest of them to come to us."

"Then it's agreed. Inform me when your unit is ready to mobilize, and I'll send a few-"

EEYAAHHHHHH!

"What the hell was that?" Feltstone said as he snatched his piece from his belt.

More screams from all around them, quickly followed by several gunshots.

"They're here!" Tanner shouted, spinning to face the door. Krik, Pounder, and several others withdrew their pistols from their leg holsters and moved to form a protective barricade in front of their boss, while Feltstone's men did the same. So much for unarmed.

The gunfire rose to a frantic crescendo before stopping abruptly. A wet moan was uttered through the resonant ringing in the ears of all those present.

"Did-did they get them?" Krik asked no one in particular.

"Shh," someone else hissed.

A large section of the back wall of the warehouse crumbled inward as the marked police car bulldozed its way inside. Lengthways. The smell of hot rubber choked the air as the vehicle screeched to a halt against a stack of octopus-stamped crates, leaving a trail of chalky dust that was slow to settle.

Both factions opened fire indiscriminately on the hole in the wall until Tanner screamed at everyone to stop shooting at the empty space. Waving away a miasma of gunpowder, he looked first to Feltstone, then nodded to the vehicle. Feltstone raised a finger to the nearest of his men and directed him to investigate. The undercover was unable to disguise his reluctance as he approached the dented squad car, reaching a hand ahead of him as if he were about to test for fire on the other side of a metal door.

"Empty," he said, after eventually plucking up the courage to peer inside.

As his men gathered themselves and refocused on the darkness that was seeping inside, Feltstone addressed the unseen: "Whoever's out there, you're under arrest. Not only have you caused substantial damage to private property, but you're interrupting official SCPD business. This is your one chance to show yourself willingly, or we will come to you with extreme prejudice."

The only answer was the rising wind as it licked a loose brick from the edge of the opening like a child's tooth.

"Warren. Pete. You two take point," he ordered, "the rest of you be ready for anything."

Two of the officers peeled away from the rest of the group and made their way towards the opening, weapons raised. Tanner and his men were already backing up towards the main doors, intending to slip out before whatever was outside decided to come in. The Tricks must have been tailing them the entire time. How else could any of this be explained? He was reluctant to believe Feltstone was in on it—he could practically see the capillaries rupturing across his eyes from here. They were almost at the door by the time the men reached the opening and were yanked into the night.

"GO!" Feltstone roared as the rest of his forces rushed the opening.

The tangled shapes of what were formerly known as Warren and Pete exploded from the void and bowled through the approaching crowd. Feltstone threw himself to the chipped concrete floor to avoid taking a flailing shin to the head, his gun discharging and punching a hole in a foul-smelling crate that leaked off-pink ice cubes. The dead men tumbled across the ground before stopping. "GET OUT!" he screamed. Pounder and Sonny were first to the door, slamming against it and rebounding into those directly behind. Sonny was back on his feet and kicking at the bent handle before Pounder had gathered herself, screaming something about her knee. The nearest thug pulled her to her feet while Sonny continued kicking in futility.

"It's no good, Boss. The handle's jammed from the outside."

"Out the way." Tanner raised his Glock and fired three shots in quick succession, each ricocheting with a drawn-out wail.

"Stop!" Sonny screamed. He put a hand on his left shoulder and pulled away to see blood. There was no time for pause as a guttural growl came from behind, and a towering figure stepped into the piss-coloured light of the warehouse.

"Bandera?!" Tanner yelled. Someone was going to pay for this fuck-up when they made it out. *If we make it out*, said an unwelcome voice in the back of his mind.

The allegedly-dead hitman wore a bloodied SCPD issue uniform and a sneer on his face. Pretty ballsy for a man with at least a dozen guns pointed at him.

"You sonovabitch," Feltstone spat as he saw the word "Creek" stitched over the breast of the man's uniform. "You're going to pay for what you've done."

The figure's sneer widened. "Bring it."

Once again, the warehouse exploded into a compressed cube of chaos. Crates shattered and shelving toppled as the figure shrugged off the metal bees and ran at the nearest plain-clothed officer. He drove his fist into the man's stomach with enough force to send him cartwheeling through the air and into the strip light overhead, raining viscera and glass down on anyone unfortunate enough to be below. The attacker took a well-placed shot to the knee and stumbled momentarily as he reached the next gunman, wrapping his arms around him and pulling him down to the floor.

Chief Feltstone had crawled behind the nearest stack of crates and watched in horror as the assailant used his officer as a meat shield while rising to his feet. The man jerked violently with each round that punched wetly into his back and head while the man holding him advanced.

Further back, the Mole was at the door, searching desperately for a weak point while the rest of his men stood behind him, firing at the man in the uniform. Had he been paying attention, he would have noticed the bullet-riddled officer being hurled in his direction and may have been able to avoid taking the full force of the impact to his head. His neck snapped forward, face smashed against the steel door before he had a chance to utter his surprise. Someone screamed his name, but the world was too red to make out where it had come from. Several pairs of hands grabbed him roughly, and he felt himself being dragged along the floor.

Feltstone crept around the side of the crates as the monster of a man passed by, leaving a trail of disembodied limbs in his wake. All but one of his officers had been massacred. He peeked from his position just in time to see the last, Enrique, have his lower jaw torn from his face and slammed into his eyes, the teeth glimmering like a cheap tiara. Feltstone waited a moment longer before making a dash for the hole in the wall and slipping into the night.

"PUSSY!" Sonny yelled as the police Chief left them for dead. Not that he could blame him—he wanted nothing more than to be out of there. His pistol ran dry, but Krik slapped a fresh clip into his hand while continuing to fire his own. A moment later, Krik ran dry, he realised he was completely out of ammo and backed up to the corner of the warehouse, to where the others had dragged his boss. Then, he spotted something that instilled a little hope.

The stock of an AK47 protruding from a blown-apart crate.

"The crates!" he shouted to anyone still alive to hear him. Sure, the weapons inside would be sans-ammunition, but there would be several packing something far more useful with any luck. He cracked the lid of the nearest and retched as the fish stink of the contents hit him like a sock full of coins. Three others joined him in his search while Pounder, Krik, and a couple of others continued shooting.

The not-dead hitman was two-thirds of the way across the expanse of the warehouse, the sneer still glued to his face. It remained even after Sonny blew his left ear apart. Then, a dark gash bloomed along the side of the man's throat as a hot lead projectile punctured it and passed straight through. For a moment, Sonny could see straight through the wound to the smoking police cruiser on the other side, but a sloppy black substance rapidly filled the gap, solidifying as if it were rapid-set glue. Sonny fired again, clipping the guy on the upper arm and twisting his body to the side. Another guy—Yip, a young recruit with something to prove—charged the monster with a switchblade. He buried the shining steel into the man's stomach and twisted it violently to the left. The man looked down at him as if he were a baby that had just done something adorable, then grabbed Yip around the throat with one arm and lifted him off the ground. Yip's feet kicked wildly as he beat at the man's forearm with both fists.

"Sonny!" someone shouted, pulling him back from the edge of insanity.

Krik had pushed over a stack of crates, shattering them and spilling a mixture of squid, assault rifles, and hand grenades across the floor. Pounder already had a grenade primed, and Sonny took a few steps back to get out of the way. The killer dropped Yip's spasming body to the ground, his snapped neck no longer keeping his face from kissing his chest, right on top of the grenade that had just landed at his feet. Yip's body was thrown into the air as it absorbed the majority of the blast and collided with the nearest row of shelving, which tipped over and started a domino effect on the rest. The stink intensified tenfold as all manner of dead sea creatures skidded across the floor on beds of ice.

A second grenade hit the man in the chest while he was distracted, swatting at falling crates, and tossed him straight into the path of the collapsing shelving.

"GO! GET TANNER OUT OF HERE!" Krik shouted.

Pounder and one of the other heavies helped their boss to his feet. He shook them off, over the initial disorientation but still bleeding heavily from a large gash on his forehead. "Give me one of those fuckin' things," he snarled. Pounder dropped the grenade into his blood-slicked palm. He pulled the pin and threw the hot potato at the area where the hitman had been buried. A moment later, an explosion of wood, bent metal, and black powder shot in all directions. Tanner addressed Sonny: "Find him. I want his fuckin' head on my desk." Sonny nodded once, internally cursing, and started to climb the wreckage while the rest of the organization hurried across the warzone and through the gap in the wall, sidestepping the crushed and twisted bodies of several of the men that'd been posted outside on their way back to their vehicles.

The door they'd initially entered through was indeed sabotaged—the handle bent into a pretzel—and the blood of the dead speckled the smashed-up cars in the parking bays. Fortunately, as they'd positioned them-

DEAD HARD

selves a little further up the street, the organization's transport remained untouched, and they piled in without looking back.

As Krik put the warehouse in the rear-view mirror, he could have sworn he'd heard a short, sharp cry.

Benjamin released the man's crushed skull and shrugged the rest of the shelving from his body. He was floored by his ability, unable to suppress the smile creeping across his face the entire time he was slaughtering them.

Although his memories of his former life were still somewhat foggy, he was pretty confident he would have never considered hurting another human being, so was it whatever had happened to him that was making him revel in the suffering of others? Or *was* it all him? A sense of retribution? Which brought another thought to mind: *Bandera*. That's what they'd called him. If they thought he was someone else, it would certainly explain why he'd been targeted in the first place. He would make sure to look into that one later.

Benjamin stalked the ruins of the fishery, searching for any survivors. He'd done a good job of ensuring the bodies remaining were well and truly expired, but several had made it away. No matter. He knew where to look for Chief Feltstone. Tanner and his crew may be a little harder to locate, given they owned the majority of the city and no doubt had multiple places they could scurry away to regroup. It would make things a little more complicated now that he knew they were coming for him, but he wasn't worried—the graphene was somehow capable of plugging the wounds he had sustained in the attack. The grenade had torn apart the left side of his uniform and charred the skin beneath, but it hadn't hurt, which was a little concerning.

Was he dead? He didn't *feel* dead. And dead people usually don't keep walking around. He wasn't sure he'd ever get the answers to the questions that were continuing to build, and perhaps he wouldn't need to. Whatever the case, he'd been gifted with the power to clean Silver

MATTHEW A. CLARKE

City up and restore it to what it always should have been, before the gangs tore it apart from the inside and the police turned a blind eye.
   If he didn't, then who would?

Chapter Eleven

This was the place.

Detective Schafter killed the ignition and stepped out onto the baked asphalt across the street from Advance Tomorrow Industries. A modern-looking complex—all square edges and large windows—it spanned several floors and wrapped around itself in a large C. Dozens of cars were parked on the other side of the wide, chain-link fence, indicating business as usual.

Schafter approached a lone security guard in a small pillbox and flashed his badge to the disinterested clerk, who looked up from his porno mag (poorly disguised inside a too-small copy of Nat Geo) just long enough to wave him through. If this was the extent of security, the place was practically begging for trouble. He noted several cameras and motion detector lights as he made his way across the lot towards the main entrance and into the lobby.

An older Mexican lady was working the reception desk beneath a large metallic sign that bore the company's name. She held up a hand as Schafter approached. He waited patiently, surveying the minimalist open space while she spoke animatedly to someone on the other end of the phone. Three corridors branched off to different areas of the building, marked 'Research and Development', 'Robotics and Processing', and 'Marketing'. Although he was fairly sure he knew which one would lead him to what he was looking for, he decided to wait to identify himself to the woman.

"Detective Schafter," he said when she finally ended the call with who he'd gathered was her boyfriend.

The receptionist opened a bottle of red nail polish, poisoning the air of the sterile environment. "How may I help you, Detective?" she asked as she began to lather her overgrown claws.

"I'd like to speak with someone from your R&D department."

"O'Laurel up to no good again? I told him selling those dodgy phones would come back to bite him in the arse."

"Uh, no. Nothing like that. I'm following up on a lead. I have a few questions I'd like to ask someone who knows a little more about the subject than myself."

"Oh, okay," she said, looking a little deflated. "Well, just don't go checking his locker, is all I'm going to say on the subject." The receptionist winked. "The corridor directly behind you, follow the yellow line until you reach the labs."

Schafter thanked the woman, wondering what on earth O'Laurel must have done to her to make her have it in for him so badly, and started down the polished poly resin corridor. The yellow line along the wall took him past several server rooms and two offices before he reached the swing doors to the labs.

Several people in white lab coats and safety glasses stood in front of workstations covered in all manner of devices Schafter couldn't begin to comprehend. He recognized a Bunsen burner from his schooldays, but everything else may as well have been from Mars. He approached the nearest worker—a heavyset man with a permanent scowl etched on his face through years of frowning—and introduced himself. The guy seemed reluctant to talk and told him he had no idea about anything to do with missing graphene before turning his back to him. The next worker told him more of the same, then muttered something under his breath about cops.

Reluctance to cooperate was nothing new to him. The people of Silver City had lost hope in the force long ago, and as much as it pained him to admit, the majority were only able to scrape by participating in questionable deeds. He was close to giving up and going back to his life of thinly veiled discontent when a young woman with dark eyes and black hair pulled back in a tight ponytail approached him.

"Excuse me. Are you a policeman?" she asked, somewhat warily.

"Detective Schafter, SCPD," he said, flashing his badge again.

"Are you here about Benny?"

"Who?"

"Sorry. Benjamin. Benjamin Hard. I called earlier and reported him missing? I haven't seen him for days, and no one else seems all that bothered."

Schafter nodded politely. Could the two things be connected? Even if they weren't, it was all he had to go on at the moment. He should be careful if she'd called it in this morning, on the off chance someone from the station decided to follow it up and spot him here when he was supposed to be off duty. "Is there anywhere quieter we can talk?"

"Sure."

Schafter followed the woman back out into the corridor and through another set of doors at the far end to what appeared to be a loading bay. Racks of shelving were home to all manner of boxes and jars. Some were marked with hazardous symbols, and others without. In the corner, next to the door they'd entered from, were several stacks of large blue barrels on pallets. "So, you say your colleague is missing, Miss. . ."

"Lang. Khoai Lang. I haven't seen Benny for two days. He's more than a colleague. We're engaged. It's totally unlike him to—"

"Calm down, Miss Lang. I'm here to help. When was the last time you saw Benjamin?"

Lang paced the floorspace, her heels clacking rhythmically. "The other day. Tuesday. Here. He stayed behind to close up. He was supposed to pick me up later that evening, but he never showed. His phone's off, and he's not at home and last night I could have sworn I'd seen someone that looked just like him wearing a police uniform but I couldn't have because that's ridiculous."

Schafter waited until he was sure she was done before saying: "Did you see anything unusual between the time you last saw him and the next day you came in?"

"I-I don't think . . . Yes. Actually." She dashed past him to the stacked barrels. "Several of these containers were broken, and the lock was missing from this shutter," she pointed. "It looked like someone had tried to clear up but left an awful mess across the floor. Dig, my manager, thought it may have been a robbery, but we ran through inventory, and nothing was missing. Except for Benny, that is."

As Schafter looked the woman in the eye, she lowered hers to the floor.

"I'm not like the others," he said gently. "I know you might find that hard to believe, but I want to get to the bottom of this. I'm here because I'm investigating the murder of a police officer."

"Oh, I'm sorry."

"Don't be. The guy was an asshole."

Lang's eyes lifted, and the sides of her lips curled slightly.

"He'd been into shady shit for as long as I'd known him, but I can't ignore the fact a murder was committed. I was led here by a substance we recovered from the scene. Graphene."

Lang perked up. "That's what's in these drums! Wait . . . are you suggesting Benny has something to do with the death of a police officer? But . . . why would he have taken graphene powder from here? There's nothing he could have done with it outside of the labs."

"I'm not suggesting anything, Miss Lang, simply following up on leads. Now, you say you haven't seen him for a few days. Do you know where he lives? Have you been by his house?"

Lang nodded sharply. "He wasn't there. I waited there until it got dark, but he never showed."

Schafter pulled a notebook from the inside of his jacket. "Could you give me his address?"

Schafter left the labs and drove straight to the address provided by the Vietnamese woman. He'd performed a quick sweep of the loading bay but could see nothing that would be of any use to him or his investigation without forensics. Still, he got the feeling he was onto something. The 'what' was another matter entirely, but it couldn't be a coincidence that one of the scientists from the Advance Tomorrow R&D department went missing at the approximate time of Creek's death, nor that there had been signs of a struggle (aside from a few knocked barrels). But what connected the two? Was it possible that Benjamin Hard had also crossed paths with Tanner's gang? To what end?

Okay, so perhaps all he really had were more questions than answers, but he'd bet his left nut it was a hell of a lot more than anyone back at the station had in relation to the case. Creek was likely already six feet under, the case file buried, burned, or 'misplaced'. He didn't feel like he owed the man anything, but murder was murder. He'd seen countless cases closed without resolution, but not this one, he promised himself as he weaved around a hunchbacked hag pushing a bottle-filled trolley across the road.

Hard's neighbourhood had once been home to the theatre district, but the place had gone to shit since the fall. The exterior of many of the run-down auditoriums were still clad in bold lettering advertising long-forgotten movies and plays. Crumbling ticket booths and cracked windows covered in angular orange graffiti bore all the hallmarks of the crack-babies. Schafter had been wondering where they'd taken to hiding out, and it seems he'd finally found his answer. The crack-babies were Silver City personified, yet even with the city in the state it was, they still chose to live in the shadows. Before the city was regarded as entirely irredeemable, they were the subject of spooky stories told around campfires, warnings told by the parents of naughty children at bedtime, and creepypastas on the internet. When the crack manufacturing process had been perfected, those that were able to function while on it found themselves capable of incredible speed. Most were driven mad, but few excelled, becoming violent but intelligent anomalies.

Schafter shuddered at the thought of what may lay inside those theatres as he pulled up outside the address Lang had given him. He was mildly surprised to find the front door was missing. For a second, he considered calling it in, but then he caught himself, remembered who it was he worked for. Still, he'd brought his radio with him in case of emergencies, so he switched it on before crossing the neat lawn, should he need to use it in a hurry. "Anyone home?" he called into the gloomy hallway, one hand on his holster. He stepped inside the building and was hit by an overwhelming metallic tang.

MATTHEW A. CLARKE

It wasn't blood, though. He was well acquainted with the smell of blood. No, this was something else. Similar, but different.

The hallway was fairly neat, with a small telephone table and a photograph of the woman from the labs and a rugged-looking man hung at a slight angle on the wall. He called out again before progressing any further, identifying himself. He drew his pistol and released the safety—if there was anyone home, they were unwilling to make themselves known. There was a good chance Hard would not be friendly if he did come face to face with him. The ground floor was clear, and Schafter once again identified himself as he mounted the stairs to the upper floor with his pistol aimed at the top. A small landing led to an office with nothing but a wooden desk. To the right was a bedroom with an unmade double bed and a cheap wardrobe with a few items of clothing on plastic hangers, and to the left was a bathroom. Schafter could see a bathtub with a large crack running the length of the porcelain from the hall but had to kick the door fully open with the tip of his shoe to see the rest of the space. Above the sink and to the left of a toilet was a square of bare plaster where the medicine cabinet had been fixed before it was torn from the wall and emptied onto the floor.

It was then that he noticed the smears. All across the sides of the shattered glass cabinet doors was a blackish-red substance similar (potentially identical) to the one he'd had swabbed at Creek's place. He peered into the bath and saw a small pool congealed around the plughole. Holstering his weapon, he bent to probe it, before smearing it between his thumb and forefinger. It was jellified but gritty. Grimacing, he flicked it back into the bath.

So, it was likely that whoever had been at the murder scene was the same person that had been here. Only here, there was no body. The only thing out of place (aside from the front door) was the medicine cabinet, which indicated that they were familiar with the place and came here looking to treat themselves for. . ?

That's where Schafter drew a blank. Everything pointed to Benjamin Hard being the killer. But why? From what Lang told him, he seemed like a regular guy, a little boring, but he worked hard and tried to keep his nose clean. Schafter supposed he may have had a run-in

DEAD HARD

with the wrong crowd. It wouldn't take much—if anyone from the Tanner's crew saw something they liked, they would take it. Perhaps Hard had tried to resist them, or perhaps he'd been working with them all along and was using the graphene powder to produce God only knew what.

There had to be more to it.

CCCHHRRRT!

Schafter jumped at the sudden crackle from his belt. He snatched the radio and cranked the volume.

ALL AVAILABLE UNITS RETURN TO THE STATION. THE GUY FROM LAST NIGHT. HE'S HERE. WE CAN'T HOLD HIM OFF MUCH—

The call ended. The man on the other end had sounded panicked, as if he was strapped to a dinner cart and being wheeled in front of a crowd of hungry cannibals. The airwaves were flooded with static and garbled voices as officers from across the city scrambled to respond to the call. Schafter didn't know who the 'guy from last night' was, but he intended to find out.

Chapter Twelve

Schafter could smell the station before he could see it.
 Thick tentacles of black smoke twisted and churned over the grotty skyline as he rushed across the city. The scent of burning wood and metal increasing astronomically as he drew nearer and made the already foul-smelling air almost bearable. Schafter threw his junker onto the curb and stepped out before the engine died. He had to catch himself and make a conscious effort to take the keys and lock the car unless he wanted it to disappear as soon as his back was turned.
 The detective had stopped a safe distance down the street from the blaze, and although it didn't appear to be as bad as he'd initially suspected, it was catching fast. Windows on the first floor at the front of the building blew outward and forced the officers that had been approaching to shield their faces and take a few steps back. Several SCPD vehicles had formed a semi-circle across the street, where the second line of officers were leaning over their squad cars with weapons trained on the only exit of the dying building. Schafter broke into a jog. "What the hell's going on?" he shouted. The nearest cop, Lime, mouthed a response that was drowned out by the crackling flames. Something deep inside the building popped, and then exploded, causing several of the gunmen to flinch.
 "GET THE FUCK OUT OF HERE, SCHAFTER!" Lime screamed.
 "FUCK THAT. WHAT'S GOING ON?!"
 Lime ducked behind the door of his cruiser as another small explosion rang out, and a snake's tongue of fire

darted between the open entrance doors. "I SAID GO! YOU'RE A FUCKING LIABILITY."

Schafter watched through the blown-out windows as the fire forged a path through the ceiling and the hardwood of the floor above. The windows of the upper floor cracked instantly, then exploded outward a second before a flaming shape burst forth and rolled to a stop in front of the squad cars. A female officer produced a small extinguisher from the boot of her squad car as the flaming object jerked upright with eerie rigidity. Schafter could just about make out the top of the man's skull as his skin blistered and peeled away from the bone like overdone crackling. His lips dripped to his blackened lap in a mess of bubbling gunk. A moment later, the officer with the extinguisher reached him and doused the man in a cloud of white powder that did nothing to save him from death.

Schafter peeled away from the cover while Lime was distracted and took several steps toward the burning effigy of corruption before the heat became too much to bear. His instincts were telling him he should be doing something more to help those that may still be inside the building, but the bitter smell of his nasal hair singeing convinced him otherwise. Ignoring calls for him to back away, he raised a hand to shield his eyes from the heat and peered into the crackling bowels of the lobby.

The heavy oak reception desk was now half the size, and several small black mounds that may or may not have been people were just about visible further in. Schafter was forced backwards as the ceiling gave way, and a fresh surge of heat funnelled out the door.

"WE'VE GOT THE BUILDING SURROUNDED," shouted a wavering voice from behind one of the parked cars across the street. "YOU'RE DEAD EITHER WAY."

Schafter touched a hand to his crisped eyebrows on his way back to Lime. "Who's in there? I have a right to know!"

Lime pulled his eyes from the fire just long enough to shoot Schafter a dirty look. Schafter remained impassive until Lime got the hint that he wasn't going anywhere. "A tweaker, or some shit," he grunted.

"What happened last night?"

This question beseeched Lime's full attention. His eyes narrowed. "What?"

DEAD HARD

"On the radio. They said it was 'the same guy as last night.'"

"Forget it," he said, facing the building.

Before Schafter could press the matter further, a convoy of battered firetrucks rounded the junction at the other end of the street, swerving and clipping several parked cars before grinding to a halt outside the police station. Schafter had little doubt the crew were either drunk or high, like the majority of the city's residents, a theory that was confirmed the moment the driver at the front of the three exited the vehicle and an uncapped hypodermic needle clattered to the floor alongside her soot-stained boots. Another dozen or so suited firefighters emerged from the vehicles and the ladder atop the first began raising steadily, stopping once it was level with the third-floor window—the only floor that wasn't currently drowning in hellfire. Only then did Schafter notice the panicked faces pressed against the pane, fists pounding soundlessly under the cacophony of chaos. A smouldering hand broke through and was promptly lacerated by the jagged glass, spraying arterial blood across the visor of the approaching firefighter, who smeared it away with a gloved hand.

"GETMETHEFUCKOUTOFHERE!" the bleeding man shouted as the firefighter broke the rest of the glass with a small metallic tool and dragged him out and onto the ladder. A second smoke-stained body followed shortly after, throwing himself from the window and misjudging the gap between the sill and the ladder. He grasped at the air before plummeting to the ground behind the firetruck. A third figure emerged as the bleeding cop reached the bottom of the ladder and collapsed onto the ground, his lacerated hand continuing to pump blood. Two of the other firefighters dragged him away from the immediate area while others approached the front doors of the building with a thick, serpent-like hose. The firefighter at the front raised a hand, and a fierce torrent of water blasted throughout the lobby.

Schafter had backed away a little by the time the third person had descended the ladder and reached the ground floor. Ash was drifting to the floor like the first snow of winter; fluffy, light, and innocent. Two uniformed officers, one male and one female, made a beeline through it, wrapped their arms around the coughing figure, and helped him toward the rest of the officers. "Get me out of here," he sputtered.

Chief Feltstone. If anyone were obliged to tell Schafter what was going on, it was him.

However, before he could get anywhere near, there was a resounding crash from somewhere above. A large section of the brick wall surrounding the broken window erupted outward, toppling the firefighter from the ladder and sending them reeling onto the roof of the firetruck on their way to the ground. Schafter was entirely unprepared for what he saw standing in place of the window.

A large man in a SCPD uniform and a riot mask was sneering down at them. His lower legs were ablaze, blistering pink flesh peeked through layers of blackened skin, but he either didn't seem to be aware, or he was some kind of crazed masochist that was enjoying it.

To Schafter's surprise, the police on the ground opened fire without hesitation, emptying entire clips in the direction of the imposing officer. Somehow, none of them seemed to hit their intended target, as the man simply remained where he was without reacting. The thick, death-black smoke cleared momentarily, revealing the face of the man on the third floor and the reason the SCPD were firing on what could have been one of their own.

Schafter pulled the photograph he'd snatched on the way out of Hard's residence and held it up to the man. It was a match.

The firetruck shook as the man leapt from the gaping hole and charged down the ladder, the flames from his burning legs streaming behind him comically as if he were running super-speed. But there was nothing amusing about the situation whatsoever as his fellow officers commenced the second round of shooting. There was little room for doubt of the shots hitting their intended target this time—Schafter could see small holes blooming across the man's torso as he leapt from the ladder and landed next to one of the men that had escaped the burning building. Hard/Bandera picked the disorientated man up and hurled him lengthways over the bonnet of the cruiser and into two of the closest police officers, swiping them off their feet.

Further down the road, an unmarked car sped way. Schafter could see the shrinking whites of Chief Feltstone's eyes as he watched the chaos from the rear windscreen. The factitious officer seemed to notice this too, and slipped behind a firetruck as if to give chase. A moment later, an axe appeared, followed by the ruined body of a firefighter, the glass of their mask caved in and face split in two. Schafter drew his weapon but kept a safe distance as several of the uniformed officers approached the truck, screaming at the attacker to show themselves. As Lime, the nearest officer, closed in, a flash of silver preceded his hands clattering to the floor, pistol still gripped firmly between them. His scream turned to wet gurgling as the top third of his head was lopped off by a second swoop of the axe.

The attacker stepped back into view, his filth-encrusted forehead several inches from the barrel of the second officer. Had she been more focused on her impending death instead of her partner's collapsing corpse, she may have been able to squeeze the trigger in time to save her own life. Schafter flinched as the sharp end of the fire axe was swept upward with enough force to tear the woman's lower jaw from her face and split her features in two. Her eyes bulged, perhaps realizing she was looking down either side of the street at once before she faceplanted the ground. Hard raised the axe overhead and hurled it at a third officer. It missed by a fraction of an inch. Before the officer could take any satisfaction in being alive, Hard was on him. He gripped both of his wrists and snapped them backwards like wishbones before headbutting the man hard enough to create a deep crater in his forehead. The attacker released him, and the man slumped across the bonnet of his cruiser, mouth flapping like a stabbed eel. The attacker closed in on a fourth officer as another wave of gunfire came at him, his uniform opening in small petals in several places. He reached another officer, Philpot, a coffin dodger far-too-old-for-this-shit but desperate for the paltry wage to nourish his grandson's debilitating meth addiction. Schafter saw what he took to be relief in Philpot's eyes in the split second it took the bullet sponge to tear his head from his shoulders and smash it through the vehicle's windscreen.

A suffocating surge of heat heralded a large section of the blazing building peeling away. Boiling rubble cascaded across the firetrucks and filled in the potholed ruin of a road. Several of the more sizeable chunks slammed into the side of the lined vehicles, followed by a thick blanket of noxious smoke. Someone barked, "RETREAT!"

Upon hearing the voice, Schafter realized he was already moving away from the carnage—his body taking perfunctory steps of self-preservation. Several panic-stricken faces emerged from the smog, deciding to continue their retreat to the end of the road and out of sight. Schafter followed at a steady stroll while looking over his shoulder, continually wiping sweat from his eyes.

The street-level smog was beginning to disperse. Two officers remained behind, approaching a sizeable mound of smouldering brickwork. No sign of the attacker. A male officer, a bald, heavyset man, was trying to pull the other, Lieutenant Kerry, away. Alice Kerry was an old friend of Schafter's father. When Felix joined the force, he'd been disappointed to discover she'd become as bent as the rest of them. Still, she'd always been good to him, so he imagined he would try and give her an out, should he ever manage to bring the department back to a respectable standing in society. The Lieutenant released one hand from the butt of her service pistol and began to roll some of the smaller chunks away from the steaming pile, ignoring the pleas of the man behind her.

The rubble shifted outward as the man-that-should-certainly-be-dead began to push himself out. Kerry took a step back and fired, blasting the top of the pinkie and middle finger off the hand that appeared. His other hand followed swiftly, a fist that bent the weapon on impact and left Lieutenant Kerry's fingers a twisted mess. The bald cop ran without looking back. Schafter almost turned and began to make his way back to her but stopped himself, realizing the futility of the situation before he'd even taken two steps.

Hard towered over the woman, ash and brick dust tumbling from his shoulders. Kerry didn't try and run. Instead, she swung with her good hand and delivered a solid strike to the middle of Hard's chest. He stared her down as the blows continued, toying with her like a coyote with a wounded squirrel. Schafter adopted a shooting stance, steadied his breath to the best of his ability. It was time to see if all those hours at the range would pay off. He took one final breath, then gently squeezed the trigger.

DEAD HARD

Kerry remained on her feet just long enough for Schafter to register the small hole that had appeared over her temple. He was able to fire twice more before the guilt hit him, but his aim was off, and the bullets thudded into the side of a firetruck further down the road with a hollow *twang*. Hard turned to face him, his face caked in sweat and ash.

Realising he was now alone, that he'd just killed a superior and now held the full attention of the seemingly unstoppable force in front of him, Schafter's body took the reins once again.

He ran.

Chapter Thirteen

Graham Krik tossed the joint roach into the gutter and slipped back inside the dry cleaners, scowling at the tittering of the young Asian women that ran the place. He told himself they were snickering at whatever was being said on the mumbling TV behind the counter, but knew deep down that was likely not the case. Deciding to make a quick pit-stop to the bathroom to check himself in the mirror, he slipped inside and shut the door behind him quietly, unwilling to draw attention until he could be certain his makeup was still intact, that he hadn't been walking around with those unsightly abominations on his cheeks on show to the world.

Krik paused, sensing he was not alone. He stopped mid-stride on his way to the smeared mirror above the row of yellowed sinks on the left wall and turned his attention to the stalls. Two were vacant. The third, at the far end, was closed. Krik considered leaving but was hesitant to without checking himself over first.

Whispers: "*Don't be like that. You know I miss you. It's just I've been real busy with work, and I can't always talk around these guys. They—*"

A slight pause, as if interrupted, then, "*—No. We look out for each other. They wouldn't understand. They can't hear you like I can.*"

"You mean we can't hear her at all, you freak," Krik said.

Silence.

"I thought you'd gotten over that years ago."

There was a strangled gurgle of water as the toilet flushed. Then, Pounder stepped outside, walked to the sink across from her, and washed her hands without looking over. "Please, don't tell the others."

"You need help, Sam. If you let that goody-two-shoes bitch get into your head—"

"Don't you *dare* call her . . . what in the fucks name have you done to your face now?!" Pounder's expression snapped from embarrassed to shocked amusement the instant she set eyes on Krik. "Do you realise how ridiculous you look right now?"

Krik sighed through his nose. "Don't change the subject. You told us you'd had that bone removed and buried at her memorial site."

"I did."

"Well, clearly you didn't because you're still talking to it." Krik moved to the sink next to Pounder and checked himself over in the mirror. He'd been in two minds about the idea when he'd read about it on an online Chelsea-smile support group, but the Joker-style disguise had been perfect for his situation. Sure, he was much shorter than the iconic villain, but he'd even gone so far as to dye his hair green and wear a purple suit.

The makeup was flawless.

"So maybe I missed a piece of her skull in my lower calf," Pounder said, shrugging, "big deal. I couldn't bring myself to remove her entirely. That grenade was meant for me. The only reason I made it back from Iraq was because she threw herself on it."

Krik straightened his lime-green tie. "I know, Sam. We've heard the story a million times. There's no way she's been communicating with you through her bone fragments. Survivor's guilt is a thing, you know? You have to accept that she's gone. Move past the trauma."

Pounder slicked back her hair in the mirror then looked Krik up and down. "You're one to talk."

"We've all got our crosses to bear, I guess."

The bathroom door opened just enough to allow a disembodied voice to slip through: "What're you two doing? Boss is waiting, and he's *pissed*."

"He's always pissed," Krik muttered as the door closed. "Let's go. I don't fancy ending up like Red."

Krik ignored the skinny shape that flittered throughout the shadows as they proceeded to the end of the corridor and down the stairs. The door to Tanner's office opened before Krik could wrap his fingers around the handle. Tanner raised a finger but didn't look up from his cigar, the desk phone clasped in a pudgy hand, so Krik used the moment to check behind the door. He knew the Skinny Man had opened it, but he wasn't there now. Krik had no idea how the man moved with the agility and stealth of a panther, but he got the impression he wouldn't believe it if he did.

Tanner slammed the phone back on its cradle. He removed his glasses and rubbed his temples with his pinky fingers. "Sit."

Krik took a seat to the left of Sonny, taking note of the way his legs were jogging erratically, leaving Pounder to take the one on the right (which, judging by the dark, tackiness of the floor beneath it, had been the one in which Red had been sitting when the boss offed him). They sat in silence for several moments, unwilling to attract the Mole's attention until they could gauge where he was on the fury scale—screaming angry, to straight-up-murdering-your-own-men angry.

The door opened, and a pale, sweaty-looking man poked his head in. "You wanted to see me, boss?"

"Blue. Close your eyes and count to three."

"Boss?"

"Do it."

Like the others, Blue knew better than to argue.

"Good. Okay, one," the Mole leaned back and slid the top drawer of his desk open. "Two." He produced a revolver and pointed it at the man's face. On "Three", the man hit the floor and was promptly dragged out of the room by the Skinny Man before Sonny, Pounder, and Krik could even process what had just happened.

Tanner took a deep breath before gently setting the revolver back in the drawer and easing it shut. "As much as I wanted that bullet to be for one of you three fuck-ups, I need you. It pains me to admit—truly, it does—that you're my best guys. Aside from the Skinny Man, of course, but who's to say what's going on in his mind at any moment?" He tossed his cigar butt over his shoulder. It didn't hit the floor.

"Perhaps I shouldn't have been so hasty in offing Red. What have I got left? A baby-faced idiot," he nodded at Sonny, "a schizophrenic," to Pounder, "and a fucking clown. For fuck's sake, man, what are you even supposed to be?"

"The Joker, Sir. Like the movie."

MATTHEW A. CLARKE

"I hate that fucking movie. Thinking about it makes my trigger finger itchy."

Krik shifted in his seat.

"The three of you are going to deliver Bandera's head to me personally. I want to wake up next to it to put my morning wood to good use."

Pounder failed to disguise the gag that forced its way up her throat. Although she'd been to the Mole's mansion several times, she'd only seen the inside of his bedroom once. Once was enough. He made no attempt to disguise his passion for a good skull-fucking, but her morbid curiosity had made her want to see it for herself. Sure enough, the room had been littered with decapitated heads in various states of decay.

Sonny cleared his throat. "Boss, that wasn't Bandera. Bandera is dead."

"Well, perhaps you could try telling him that? Because he seems to think it's pretty fucking funny to destroy one of my warehouses and half the fucking SCPD in the process. I don't want to hear any excuses. Just get it done."

"With respect, boss," Pounder started, "you saw how strong he was. I've seen nothing like it. It's like he's on some *super* crack or something." She looked to the others for support, "How many bullets did he take?"

Tanner adjusted the bandage wrapped around his forehead. "I may have taken a blow to the noodle, but I haven't forgotten what I saw, you crazy bitch. Which is precisely why I've called in the Silvers. I need you three to take the van and meet them surface-side. Get them over to the warehouse, and—"

The phone started ringing. Tanner let it ring a moment longer, then answered. "This better be good."

"The fucking *Silvers?*" Krik whispered barely audibly. "I don't want to get within a mile of those freaks."

"At least they're on our side," Sonny muttered.

"Are they, though?"

"They might be nuts, but they ain't *that* far gone."

The trio flinched as Tanner slammed the phone back on the cradle. Fresh beads of perspiration dripped from his bandage. "He's hit the police station. Wiped out a load more of Feltstone's guys. Burnt half the building to the fucking ground."

"There's no way he's working alone," Pounder said. "The things we saw . . . it just isn't possible. It *has* to be some kind of a trick. Bullet-proof Kevlar, hidden pyrotechnics. I don't fuckin' know."

"Whatever the case, if he's got the balls to hit the SCPD like it's just a fucking daycare, we have to assume this place is compromised. Or will be soon." Tanner pushed his chair away from his desk and stood. A lit cigar was wedged between his chapped lips by an unseen hand. "I'm going to pull back to the estate until we get eyes on him. Krik, Pounder, you two meet the Silvers. Get them to the warehouse, the SCPD, wherever the fuck else you think might be useful. Sonny, I need you to get over the city boundary. Pay a visit to the Tricks. Find out if Bandera's gone rogue, or what the fuck is going on." Tanner held his hand up, anticipating Sonny's response. "It don't feel right—if he were working for them, they would have launched a full-scale attack the moment we saw it was one of theirs. You'll be fine."

Sonny scratched the back of his neck. "Sure thing, Boss . . . but can't we just call them? Seems like it would be a whole lot safer."

"No. No good. I need boots on the ground. You gotta speak to Black Eye, face-to-face, so you can look him in the eye and see if he's telling the truth."

Sonny cursed his flawless ability to look people in the eye and see if they were telling the truth, but on the other hand, it was likely that very talent that was the only thing keeping him alive right now. He swallowed his displeasure at the situation he found himself in. "Boss."

"Good. Then it is decided. Keep your phones on. I may need you back at the estate at short notice, but I don't want to hear from you unless it's to deliver me a fucking head. Got it?"

"Boss," they muttered in unison.

"Good. Now get the *fuck* outta my office."

Chapter Fourteen

Felix Schafter had done a lot of thinking between the time it had taken him to look up Lang's address and the drive over, yet he was still no closer to knowing what he was going to say to her. The things he had seen . . . it shouldn't have been possible. There were too many things he needed to ask her before he could draw any firm conclusions, too many coincidences. Not only was the man that had attacked the station wearing an official SCPD uniform, but he was leaking the same black-and-red blood he'd seen at both Creek's place and Hard's residence. But why would a lab geek like Benjamin Hard be on a rampage? And why did he drop off the radar in the first place?

"Detective Schafter?"

Schafter flinched at the sudden interruption. He looked to his right to see Khoai Lang standing next to him, a large bag of groceries in either hand. How long had he been standing there?

Lang's eyebrows pinched the bridge of her nose. "Is everything okay? Is . . . is Benny okay?"

Schafter pulled himself together. "Sorry. Uh, yes. As far as I'm aware."

Lang stared a moment longer. "Right. Okay. Well, seeing as you're here, would you mind grabbing my keys?" she said, nodding to the front pocket of her white jeans.

Schafter instead offered to take the bags, feeling it would be inappropriate to slide a hand into the pocket of the fiancée of the man he was investigating.

"Thank you." Lang smiled and led him up the short path to the house. "Would you like a drink? Tea? Coffee?" she asked once he'd set the bags down on the kitchen counter.

"No, thank you. I just have a couple of follow-up questions, if you wouldn't mind giving me a few minutes of your time."

"Of course. Anything you think might help." Lang's eyes flitted to Schafter's left ear. "Would you like a tissue?" After a moment, "You're bleeding."

Schafter touched a hand to his ear and inspected the dull substance that flaked between his thumb and forefinger. "It's not mine."

Lang paused in front of an open cupboard, a cereal box in one hand.

"It's not Benjamin's, either."

Lang shelved the box and turned to face him. "What's going on, Detective? I come home to find you standing outside my house, stinking of bonfire and covered in someone else's blood. I can tell there's something you want to say to me, so please put us both out of our misery and just say it."

Schafter sighed as if working himself up to something. "You said you thought you saw Benjamin that night, outside your house? Wearing an SCPD uniform?"

Lang nodded. "Have you found him? Please tell me you've found him."

"I'm not sure. What I'm going to ask you next may seem a little peculiar, but please try and answer truthfully." Lang nodded once more, lips pursed. "Can you think of any reason why Hard might have a grudge against the SCPD?"

"What is this about? Has something happened!?"

"Is there somewhere we can sit down and continue this conversation?"

Lang took Schafter through to a front room. A double sofa and a lone end table faced a small television below a framed painting of several jellyfish floating along a city street. Lang sat in charged silence while Schafter filled her in on the events prior to his visit, her face growing considerably paler by the minute until finally, Schafter finished his piece, and she said: "But that doesn't make any sense!"

Schafter wrung his fingers. "I know, which is why I'm here. I'm hoping you might have some idea of Benjamin's current whereabouts." Lang's eye's widened, but before she could voice her protest, Schafter continued:

"I'm not suggesting you're keeping anything from me, Ms Lang. I'm simply asking you to think if there could be anywhere he might be hiding out right now."

Lang's lips trembled as her elbows rose to her knees. She buried her face in her hands to hide the tears that were starting to break the surface. Schafter watched for a moment as her shoulders hitched gently. "I'm sorry. I just don't know."

Schafter stood, already regretting telling her the unfiltered version of what had happened. "I'm sorry to bother you, Miss Lang. Please call me if you think of anything at all."

The room was lit with strobing lights as the siren of an emergency vehicle screamed past the window. Lang raised her head, her gaze settling on the painting above the television. She wiped her eyes with the sleeve of her blouse and took a stuttered breath. "There is one place that holds some sentimentality to Benny," she muttered, almost to herself. "He would always mention it when he spoke of his parents, but it closed down years before we met."

The detective followed her wet stare to the painting. He wasn't a fan of art, but it was beautiful, in a surreal sort of way. "Wilfred's Water World."

"Yes," Lang turned to face him once more, "but I don't see how that's any help to you. I'm sorry."

Schafter put his thumbs through his belt loops and said, "I'm going to level with you, Miss Lang, as I think you deserve to know. Your fiancé happens to bear a striking resemblance to one Julio Bandera, an out-of-town gangster, which could go some way to explaining his disappearance. What it *doesn't* explain is why you thought you saw him outside your window, in the same SCPD uniform I saw him wearing downtown earlier today. Now, I don't believe the SCPD know Mr Hard and Bandera are similar in appearance... I assume no one's contacted you since you reported him missing?" Lang shook her head. Schafter continued, "I think they believe the attacker to be Bandera. *I* believe that either Mr Hard is aware that he's being framed and is lying low, or he and the attacker are one and the same."

"He would never!"

"It's incredibly unlikely to be the latter," Schafter said, moving for the door. "But if it is, I need to know why."

"Let me come with you."

MATTHEW A. CLARKE

He turned.

"Please. You can't just drop a bomb like this on me and then leave!"

Schafter considered the woman's request. He had no idea what he could be getting into—it could be nothing at all—but it would be helpful having her there on the off chance they did find Hard at Water World. "Stick close to me and listen to what I tell you to do."

Lang shot up as if someone had just run a thousand-watt current through the sofa. "Yes! Wait. Hang on a minute," she rushed out to the hallway and up the stairs, "let me just grab my gun!"

Perhaps she wouldn't be in as much danger as he'd thought, he mused. After all, despite her meek demeanor, it takes a particular type of person to survive in Silver City.

Chapter Fifteen

The blacked-out crew van couldn't have been more conspicuous if it had BAD GUYS INSIDE sprayed across the windscreen. Fortunately, things like that didn't matter as long as the SCPD were under their thumb (if there were even enough of them left to *be* under the thumb). Still, knowing they were free from fear of prosecution did little to quell the unease in Pounder's stomach as Krik yanked the wheel and sent the tires crunching over a bed of scattered aggregate beneath the rusted arch of the Dapper Silver Mining site.

Watch the wires, Fern, the skull shard, whispered from Pounder's calf. She relayed the message to Krik just in time for him to swerve to the left and avoid one of the many traps.

Krik's knuckles whitened as he tightened his grip on the steering wheel. "Why haven't they disabled these fuckin' things? They knew we were coming."

"Maybe they forgot. Or maybe they thought it would be funny. I hate these assholes."

It's not too late to skip town, Sammy. You're really going to put your life in the hands of a guy that goes out in public wearing mime makeup?

"Shut up," Pounder mumbled. She had to admit, Fern did get a little irritating at times. Ever since she'd exploded, the woman had developed an irrational fear of death.

"If you gotta talk to your imaginary friend, can you at least wait until you're alone? I'm trying to concentrate here," Krik said before slamming on the brakes to avoid another tripwire.

MATTHEW A. CLARKE

They wound down the approach to the mine entrance and parked outside the 'office', only then realizing they'd both been holding their breath for the remainder of the journey. Krik left the ignition running as he leant on the horn, eager to spend as little time as possible in the immediate vicinity.

The door to the raised container was kicked open, and a naked, pissed-looking man with a throbbing erection stepped out onto the top step clasping a bloodied carving knife. "Kill the engine!" the Silver's unofficial handler yelled. Then, after Krik complied, the man's face softened. "Appreciate it, buddy. They'll be up in a minute. Keep the engine off until you're leaving though, will ya? We're trying to shoot a snuff in here."

Krik saluted the man and turned to Pounder as he stepped inside, muffled male screams started up almost instantly. "Everything about this place is messed up. I don't know why the boss even works with these guys anymore."

"They have their uses. The Silvers have saved us a hell of a lot of killing."

"I know, I know. I just prefer it when we're able to keep out of it. They give me the heebie-jeebies—CHRIST!"

Krik's piece was in his hand before he realized what he was doing. Fortunately, he was able to catch himself before he put a hole between the eyes of the smiling silver face pressed against his side window. The face retracted and joined the rest of the giggling group.

Pounder tensed at the sight of the balding child-like creatures. Their features were sharpened to an almost exaggerated degree, and their eyes were wide and sunken. However, these folk were the antithesis of their appearances; living in the depths of the old silver mines had given them heightened senses and extraordinary dexterity. (Although there was not a single one among them that was not suffering from mercury poisoning to some degree). The silvers had been a team of miners, at one point, many years ago. They'd been trapped in a cave-in and presumed dead until long after the mines closed—a cave-in that was later found to be caused intentionally. They had, in fact, been hoarding enough supplies to get them away from their mundane lives for several years. However, when the men and women lost in the supposed cave-in were spotted milling about the site by passers-by years later, it quickly became evident that they were not the people they once were. Driven to the brink of insanity by a life underground exposed to god-only-knew-what and unpredictably aggressive, the SCPD decided the Silvers were better off left to their own devices, and the missing person cases were closed. Since then, they'd been hired by Tanner on numerous occasions, whenever a job required a particularly ruthless, or ferocious approach.

Krik wound his window down an inch. "Very funny. You got me. Now get in the van."

Last chance, Samantha. Pounder smacked herself in the calf as the back door slid open and the five cave dwellers slipped inside. Their presence choked the air with an earthy, metallic tang. Krik lowered his window the rest of the way, started the van up, and started up the steep incline toward the entrance. He wanted to ask them about the tripwires but knew there was little point—they either didn't understand or care to listen to anyone other than their handler—and he had no intention whatsoever of finding out precisely what was going on back in that container. He weaved through the first two tripwires and chuckled along with the Silvers as the third and final was detonated by a grubby homeless kid chasing a squirrel up ahead. Blood and viscera fell from the sky, coating the windows.

Krik flinched as one of the females in the back licked the back of his shaven head. "Get the fuck off."

"Aw, I think she likes you," Pounder teased. "You should probably take what you can get, given your appearance."

Krik shot her a glare, then adjusted the rear-view mirror. The bald, bird-faced woman wiggled her mud-caked fingers at him daintily. It took every ounce of willpower not to lose his lunch. "Where to?" he asked after knocking the mirror to the ceiling.

"Well, he was last seen downtown. I guess we just head there and let these guys do their thing."

"Do their thing!" one of them parroted.

"Do their thing!" the rest screamed while bouncing in their seats.

"DO THEIR THING!" the smaller of the bunch screamed before sinking his teeth into the throat of the Silver sitting next to him and tearing out a large chunk of flesh.

MATTHEW A. CLARKE

"Jesus on a fuckin' pogo stick!" Pounder shouted as blood sprayed the back of her shoulders.

The Silver that had been bitten made no attempt to stem the bleeding. Instead, he simply joined the others in giggling at his misfortune.

Krik put his foot down.

Chapter Sixteen

A gargantuan octopus marked their arrival. One tentacle was raised in what was assumedly meant to be a greeting, while others snaked about the air above the rows of vacant spaces in the vast expanse of the car park. The mouth of a giant epoxy-coated killer whale sculpture served as the attraction's main entrance. Schafter looked up at the oversized, cartoonish eyes and shuddered at the memory of the last time he'd been inside. The last time *anyone* set foot inside.

Wilfred's Water World had been one of the main draws for bringing out-of-towners to Silver City before the owner was outed as a paedophile by a group of concerned parents. As part of the subsequent investigation, the staff had also been brought in for questioning, and a messy web of child trafficking unfurled. Schafter had been one of the officers on the scene when they came to close the place down for good. He'd never forget the sights he saw in the cages beneath the petting pools, nor what they'd been doing with the sea urchins. It transpired that one of the main reasons Wilfred's Water World had been so popular was because of the owner's willingness to pay for and 'dispose' of any unwanted children that should come his way, as well as handing out refugees (crammed in with shipments of feed) to desperate couples unwilling, or unable, to pass through the proper hurdles required for adoption. However, the staff were not aware that Wilfred was getting high on his own supply, so to speak.

"You ever come here as a kid?" Schafter asked as Lang shut the car door behind her.

"Yes. Legally, before you ask."

Schafter huffed. He wasn't used to the woman's dry humour. "I wasn't going to."

They ascended the stairs to the whale's mouth and could see from the heavy chain wrapped around the handles and the planks nailed across the door that it hadn't been accessed. Schafter led the way around the left side of the building to a delivery door at the back and found that it, too, was secure. After checking a staff-only door on the right side of the building, it seemed as if their trip had been a waste of time.

"What now?" Lang asked expectantly as if Schafter would know precisely what to do.

"Now I take you back home. I want you to stay there and call me the moment you think you see or hear anything."

"We're just giving up? I thought you said he might be in danger? You expect me to sit around at home waiting for the phone to ring?"

Schafter paused halfway down the steps and turned to face her. "Look, I know it's not ideal, but—"

A deep, dull thud cut him off. There was nothing but the empty expanse of car park around them. Only one place the noise could have come from. Lang was already up the stairs and smacking the front door with open palms. "BENNY? ARE YOU IN THERE?"

Schafter hung back a little and surveyed the exterior of the decrepit building. He hated himself for what he was about to suggest. "I think I saw a broken window on the upper floor. I may be able to access it from the roof."

"Huh?" Schafter pointed back toward the giant octopus, his finger following one of the raised tentacles that stopped just short of the entrance. "Really?"

"Unless you have a better suggestion?"

"Okay, but I'm going first."

Schafter could tell from the look of determination on the woman's face that arguing the point would be fruitless. Besides, he wasn't exactly here on official police business, was he? Just like her, he was here as a civilian, so why not let her go first? He moved aside as she bounded down the steps two at a time and hopped over the decayed fencing that had once protected the sculpture from such a thing that was about to occur.

"You coming?" Lang called after scaling the rotund body of the creature like it was something she'd done a thousand times before. By the time Schafter had made it to the top of the octopus's cartoonish beak, Lang had shuffled halfway along the tentacle. "I heard it again!" she shouted, speeding up.

"Wait!" Schafter growled, panting as he ducked beneath the lower section of one of the thing's tentacles and scrambled to the top of the head. He wrapped his hands around both sides of the tentacle and pulled himself up, using the creature's suckers as handgrips. It looked much higher than it had done from the ground, and the arm was only getting narrower the further he shuffled along on his rump. About halfway across, there was a short, sharp, cracking sound, and a thin line appeared around the circumference of the section of the tentacle directly in behind him. He felt a little woozy.

"There's too much weight on it. Get onto the roof!" Schafter ordered without taking his eyes off the crack. He remained stock-still as the tentacle vibrated between his thighs, afraid to even breathe too deeply.

After a moment, there was a slight springing sensation, and Lang said, "I made it!"

Only then did Schafter feel secure enough to raise his head to look at her. Lang waited at the flat roof's edge for him as he began shuffling along the support before he could think about how ridiculously dangerous what he was doing was. He was two feet from the end when there was another, louder crack, and the end of the tentacle dropped several inches. He heard Lang cry out for him but was too petrified to do anything but tighten his grip and close his eyes as the end of the tentacle tore away from the body and—

CRACK!

—stopped?

Schafter opened his eyes as hands pawed at his shoulders, pulling him forward. The tip of the tentacle had struck the edge of the roof, wedging it at an angle against the rest of the arm, and was all that was preventing him from several broken bones. He shifted his bodyweight forward and allowed Lang to pulled him to safety, almost dragging her to the floor as he landed.

"I guess we won't be going back that way," she said once they'd found their footing.

"Thank you," Schafter said between shallow breaths.

"You can thank me once Benny is safe."

Schafter nodded as he took in his surroundings—several vents to either side of him, and ahead, beneath a

MATTHEW A. CLARKE

Hollywood-style WILFRED'S WATER WORLD sign, a run of large windows, the middle of which was broken. He motioned for Lang to follow.

They were directly above the entrance hall. A suspended metal maintenance walkway ran the circumference of the circular room. Below, beyond the life-sized, orange graffiti-tagged sculpture of a beaming Wilfred the Killer Whale, were the ticket booths and the metal detectors. Empty vodka bottles and used condoms were strewn about the place like party streamers, a tell-tale sign of a crack-baby orgy. Schafter shuddered, unsure whether it would be preferable to find a gang of adult crack babies deeper in the building or the man he'd seen single-handedly destroying the station.

The suspended walkway creaked tiredly as Lang dropped onto it. "Need a hand?" she smirked, spreading her arms as if to catch him.

Schafter ignored her offer and lowered himself cautiously. "We should keep quiet until we know what we're dealing with here."

The place smelled of stagnant water and rot, which, he could be wrong, but smelled a little too like a crime scene for his liking. He decided not to mention this to Lang as they descended the ladder to the ground floor. They stepped cautiously around the STI minefield, through the defunct ticket barriers, and along a winding, narrow corridor flanked by glass walls of murky water. He half expected a pale, bloated face to appear at any moment.

"Woah," Lang muttered as the corridor onto a large room with what had once been home to the petting pools. Here, children could stroke starfish and poke sea anemones, posing for photos while blissfully unaware of the aquarium's second, more exclusive petting zoo directly below them.

Now, the floor space of the entire area was a gaping wound, excavated to retrieve the remains of the lost souls that were left behind when the SCPD raided the place—Wilfred had sealed himself in the room and executed the children before taking his own life. The memories, much like the floor, would never be healed. "Watch your step," Schafter said as he shimmied along the edge of the wall. Despite knowing the floor below was picked clean of all evidence of the heinous acts committed, he tried to avoid looking. Shrouded in darkness, his mind would undoubtedly project unsavoury memories.

They were faced with three doorways at the far end of the room. After spending a moment trying to decide whether they should investigate the SHARK TANK EXPERIENCE, the GLASS BOTTOM BOAT or . . . they heard what sounded like a scuffed footstep from the direction of WILFRED'S WATERSPORT ARENA. Lang tapped Schafter on the shoulder as they passed through the doorway and stopped by a metal shell-shaped sign. A large, black handprint was smeared across its surface.

Another scuff, much closer this time.

Both Schafter and Lang readied their weapons as they followed the safety railings along the side of Olly Octo's Teacup Blaster and passed beneath the winding track of Shark Tooth's Revenge.

"Shit!"

Schafter almost fired a round into a resin cut-out of a man-sized pufferfish with a top hat as the lights came on overhead and a nautical tune warbled from several sets of wall-mounted speakers.

"What's going on?" Lang shouted over the roar of Shark Tooth's Revenge as the coaster rattled along the track behind them.

"Up there!"

A large, dark shape scurried up the faux reef wall behind Pirate Pete's Booty Dive, leapt onto the track of another coaster and disappeared into the wall before either of them could get a proper bead on it.

"BENNY!" Lang cried. "IT'S ME!"

Schafter followed the woman as she ran across the amusement arcade and mounted a metal ladder on the wall beside Pirate Pete's. Without looking back, she slipped through a hatch at the top, marked STAFF ONLY with a picture of a skull and crossbones. Schafter holstered his pistol and followed.

The hatch opened up onto a low-ceilinged maintenance tunnel. The sounds of the arcade reverberated off the age-worn ductwork and guided Schafter toward the poorly lit opening at the far end. He found Lang standing just outside the opening and was about to take the lead when a ceiling-mounted cannon belched a cloud of smoke. Moments later, a pirate ship-shaped rollercoaster tore through the darkness and corkscrewed past at waist level.

"Look," Lang said once the rumbling had subsided a little.

Schafter squinted through the clearing smoke at a full-sized diorama of a gang of pirates on a sandy beach. Clumps of the black, tar-like substance were strewn about the scene. They approached cautiously, giving the tubular tracks and support arms as wide a berth as the space would allow, and mounted the small step up to the scene. What had appeared as a beach from a slight distance was in fact a thin layer of sand over a solid, yellow platform. Heavily bearded pirates watched their approach with malice etched into their painted eyes. They were eerily realistic, even up close.

"You there," Schafter said to the figure on its knees behind an oversized pirate chest, "put your hands up and turn around. Slowly."

"Benjamin?" Lang said with a slight waver.

"I SAID," Schafter began, "PUT YOUR HA—" A panel in the wall to their right opened as the next coaster roared through. The cannon fired, and a fresh blast of smoke washed over them. Schafter waved it away furiously. "Lang, no!"

Lang dropped to her knees alongside the figure and put her hands on his shoulders. He flinched, then slowly turned to face her. "No. No . . . no. What have they done to you?"

Schafter maintained his distance, waiting to see what would happen next. He had a clear shot on the man—clearly the same person he'd seen downtown—but was hesitant to take it. Although a shot to the back of the neck would be enough to floor any regular person, he'd already seen what this man was capable of and wasn't willing to risk pissing him off with Lang in such proximity.

"Go," the man said, his voice like a knife through rusted metal.

"What happened to you!? We need to get you to a hospital!"

"Lang," Schafter said.

The man wheezed. "I'm sorry. Too late for hospital. Leave me."

"No, we're not leaving you. Detective Schafter? Please, tell him."

DEAD HARD

Hard shot to his feet and spun to face Schafter, knocking Lang aside and providing a clear view of his ruined profile.

His face was severely burnt across the left side. Part of the riot helmet had melted and grafted to the charred bone peeking through curled folds of red flesh about his temple. He still wore the SCPD uniform, littered with bullet holes and burnt patches which gave glimpses of glistening white fat and taut muscle.

"DON'T MOVE," Schafter commanded while trying to hold back rising bile. The smell of cooked human flesh was not one he'd ever gotten used to.

"No! Stop!" Lang leapt to her feet and wrapped her hands around Hard's wrist, attempting to pull him backwards, but it was as if he was made of concrete.

"I'm not here to hurt you, Benjamin. Just don't come any closer."

"Listen to him," Lang urged.

Hard sneered and took a step forward. "SCPD," he rasped.

"I don't want to shoot you, Hard. We're here to help."

"Benny, please! He's a friend!"

Hard took another step closer. His hands balled into fists.

Behind Schafter, the doors snapped open, and the cannon fired. He heard the coaster rattle past, then the door closing behind it. He then heard a gunshot and, for a moment, thought he'd put a little too much pressure on his trigger, expected the beast of a man to tear through the smoke, but he was shocked at what he saw when it cleared.

Lang stood alongside Hard, her pistol pressed to her temple. Hard's attention was back on her. The sneer dropped from his face. "Don't do it, Benny. This isn't you."

Hard looked to Schafter, then back to Lang. "They need to die."

"Not him. Please. He's different. He's been helping me."

"Mr Hard," Schafter said, hoping he sounded calmer than he felt, "I understand your frustration with the SCPD, but I assure you, I am on your side."

"See? He's not lying, Ben. What happened to you?"

"I don't know," Hard sighed, his shoulders falling just enough for Schafter to lower his guard. He lowered his weapon to the floor, but maintained a firm grip on it. "I think they killed me."

"Who tried to kill you?" Lang asked.

"Bad people. Bad people that are friends of *his*," Hard spat.

"I assure you, they're no friends of mine."

Hard growled. Lang squeezed his arm. "Please. Try and focus on me." His leer softened a little. "I'm going to get you help. Wh . . . I can't find your pulse."

"No. Bad people are looking for me, but I will find them first."

Schafter holstered his pistol and risked a step forward. "Hard . . . Benjamin. I can't let you do that. Listen to your fiancée. I can take you somewhere safe, get you looked at, and see if we can't work something out from there. No good will come of continuing this rampage."

Hard pulled free of Lang's grip and approached Schafter, his vesicated skin rubbing and crackling sickeningly with each step. Schafter fought the fight or flight instinct screaming at him to do something, anything, to neutralize the situation. Hard stopped within arm's reach of the detective. His eyes were milky pearls floating in an inky lake.

"It's too late for me," he whisper-rasped.

Schafter understood then that he was right. He was talking to a dead man. However absurd that sounded, there was no other explanation.

"I need to understand what happened to you," the detective said. "And why you did what you did downtown."

"They said I'm Bandera. They shot me, I think," he pointed to the hole in his forehead that had crusted over with a black, brittle substance, then spread his arms as if to display the ruined visage of his body, and the speckled black substance that covered him.

"Benny," Lang said as she put herself between the two, her wide eyes flitting anxiously from one man to the other.

"It's okay," the detective said. Then, to Hard, "I believe you were mistaken for a hitman of a rival gang. Julio Bandera." Hard's gaze didn't waver. Schafter cast a hand at the dried wounds across Hard's body. "We recovered some of this from a crime scene and ran some tests. It's graphene. Graphene mixed with haemoglobin. Didn't make much sense at the time. Makes even less now, if I'm honest."

Hard produced a noise which could have been a snort, then turned and made to leave.

"Where are you going?" Lang cried.

"Finish what I started."

"This isn't you! Don't leave me again!"

"Stop, big guy," Schafter said authoritatively. "You know I can't let you leave."

Hard stopped for long enough to say, "I know you can't stop me."

Schafter drew his gun and whipped it at the target in a smooth, well-practiced motion. "HARD."

A second passed. Two. Three. He gritted his teeth, then took his finger off the trigger.

The cannon fired, and the room filled with smoke.

Chapter Seventeen

Hard didn't have a particular destination in mind when he'd set out, only knew that he had to get away from the detective. His memories had been returning like a knee to the gut, enough to understand he'd already lost enough without anything happening to Lang. He had little interest in trying to understand what had happened to his body—it wouldn't have mattered anyway—and he certainly wasn't going to be going anywhere with Schafter. If he were to stay any longer, it would have been impossible to ignore the urge to tear his arms from his shoulders, and from there . . . well, he didn't want to think about what would've happened if he couldn't control himself. It wasn't that he *wanted* to hurt the man, but Hard was growing convinced that anger was the driving force behind his continued existence. It was constantly fighting to be at the forefront of everything he said and did.

He was better off on his own, answering to no one. The law didn't apply to dead people.

Hard had walked through the night and had been left alone for the most part, with the exception of one vagrant that looked as if he'd been preparing to pull a knife until he got a better look at him, at which point he scurried across the street and down a dark alleyway. He found himself in unfamiliar territory when the sun kissed the horizon, a warm red arc that slowly bled into the sky surrounding it, and he noticed something almost instantly that gave him pause.

Had he been in his sound mind, he may have noticed it was a little too careless of them to leave the sedan parked on the curb outside the launderette. The very same one he'd seen the suited men leave the scene of the detective's partner's house, with the busted taillight and bullet hole in the rear window.

A roiling heat rose from his fingertips to his chest as if his body were preparing itself for things to come. He crossed the street without looking, eyes laser-focused on the mob vehicle as if it might disappear the moment he turned away. The engine hissed and clicked, indicating it was still warm. Hard felt a twisting in his gut at the thought of what would happen to him once this was all over. Where would a dead man fit into a world tailored to the needs of the living?

The launderette was a substantial building, built across one floor. Shades were drawn across the windows, and a padlocked fire escape door on one side. There was a collapsed office block to the left and an expanse of vacant lots with *NEW BUSINESS COMING SOON!* signs which had likely been there for longer than he'd been alive, judging by their condition. He performed a sweep of the immediate area and saw no other sign of life. A black van with tinted windows was parked in the alley behind the place. He punched a hole through the driver's window, and found it too, was empty. Which meant the owners of both vehicles were likely inside the building.

Hard was faced with two options: Fast and loud, or slow and loud. He opted for the first.

The front door was made of steel and secured by several deadbolts. It caved on the first kick, then came away from the frame entirely on the second. Hard ducked around the frame as a rigged shotgun expended its chamber through the open doorway—such devices were commonplace these days, especially in business establishments—and barreled down the dim corridor to the room on his left without wasting a second. It quickly became clear that the room was of no interest. A small desk, several washing machines, and tumble driers. The building remained silent as he shouldered through the door across the corridor, which exploded into a hail of splinters. Another bust, this one full of row upon row of pressed clothing on hangers. The next room looked like a small staff room with a microwave and a fridge. The room after that was a bathroom. No sign of life.

Hard stood in the corridor, rising anger (and fuck knows what else) surging through his body. They had to be here somewhere. Looking around, he noticed the far end of the corridor dropped off into the shadows. He'd almost missed it in the poor light. Hard charged through, down a set of stairs and through another door at the bottom. He found himself looking at a male and a female. Both were tied to chairs beneath a crackling bulb in the centre of the room. Their eyes widened, and they started to scream something at him through the balled-up fabric between their teeth.

Hard cast an eye over the rest of the room. The place reeked of stale smoke. A heavy walnut desk was against the far wall. Bare, except for an ornate glass ashtray, which held several crushed cigar butts. Hard approached the male in the mime-getup and tore the gag from his mouth. He took a wheezing breath and tilted his head forward, spit dripped to the concrete floor from his black-painted mouth.

"Thank fuck you're here—AGH!"

Hard grabbed a fistful of the man's slicked-back hair and yanked his head backwards.

"Tell me what this place is."

"I don't know! I was knocked out and brought h—" a cracking from somewhere within the man's upper body as Hard yanked his head back further, "AG! OKAY, PLEASE. NO MORE!"

The man was crying, and they were barely started. They were still underestimating him. He'd seen the car outside. The way the woman was dressed.

Hard waited for the mime to continue while his tears cut tracks through his face paint. With his free hand, he tore the gag from the mouth of the woman with the buzz cut.

"Fuck you," she spat after taking several breaths. "WHAT THE FUCK ARE YOU WAITING FOR, YOU INBRED PSYCHOPATHS!"

A flash of white preceded a short, sharp explosion.

Hard was blinded. Deafened. He put his hands to his eyes, assuming he must have had something thrown in them. Something tackled his legs and knocked him to the floor. Something else was on his back, while more were pulling at his wrists. Hard rolled over and felt something pop beneath his heavy frame. He shook his head and blinked several times as his vision started returning. He had no idea what he was looking at, but they looked humanoid, if not for the silver skin and the small, pointed teeth that were currently tearing ribbons of flesh from him. He swung his arms up, then smashed them back down into the concrete, splitting the skull of one and sending it into a twitching fit, and breaking the leg of the other. He then stepped over the corpse of the one that had been on his back and swung his leg against the side of the heavy desk, snapping the spine of the one chewing at his shin. Its broken body flailed behind it as Hard pulled away, its teeth still firmly buried in his bone. He reached down and tore its body from its neck, then hurled it at the one with the broken leg who was hopping toward him, laughing as if it was having the time of its life.

He turned his attention back to the two in the chairs. They were no longer bound. A fifth silver freak stood behind them with a mouthful of frayed rope.

There was a brief *crackle*, then Hard's body went rigid. Two thin wires ran from a box-like device in the female's hands to his chest. The little silver freak charged between the pair and sank its teeth into Hard's groin. Attempts to swat it away were impossible—his muscles contracted and seized by the voltage pulsing through his metal-laced body.

"Holy shit! It's working, Pounder!" the male said.

"You don't say. Shut up and get over there before I run out of juice."

The mime approached with his own device, wincing as he looked down at the creature gnashing at Hard's groin. A small arc of blue lightning crackled between the prongs on the end of the device, which the man promptly thrust into the side of Hard's neck.

He fell to his knees, shaking violently but somehow staying conscious. The silver freak tossed a flap of meat over its shoulder and worked its way up to Hard's chest.

Pounder shut her taser off, then approached the silver person and rapped it on the back of its bulbous, silver head with her knuckles. It spun to face her, teeth full of tendons and skin. "Don't kill him yet. I want to have some fun with him for what he did to the others." Then, to the mime: "Hurry up and knock him out."

DEAD HARD

"I'm trying!"

Pounder aimed her taser at Hard's forehead, then fired.

"Never known a body to be this tough to cut through. What is all this black shit?"

"His eyes are moving. Zap him again."

"Did you see that? Is he waking up?"

"Maybe. Battery's dead. Hurry up and pass me the blade so we can get his head and get out of here. The boss wants his cock in it by the end of the day."

A dull pressure on the left side of his throat. The sound of metal splitting stubbled skin.

Hard shot upright and thrust out with both arms, pushing the hunched mime off him with his forearms and sending him cartwheeling into the far wall. He swung at the woman to his left, but instead of connecting, his fist passed inches from her face through the air. She squealed as her eyes were splattered with a thick spray of black gunk, her hands going to her face as she dropped the dark items she'd been holding. Hard looked to the burnt pair of hands on the floor, then down to the stumps of his own wrists.

They'd butchered him.

The woman tripped over the corpse of one of the silver freaks, cracked her head against the concrete, and fell silent. Hard attempted to roll to the side as the one remaining Silver pounced from the desk, blood-flecked teeth gnashing through the air between them, but failed to move far enough, the pressure at his neck increasing as he pushed against it. Giggling manically, the Silver landed on his side and sank its teeth into him, shaking

MATTHEW A. CLARKE

its head like a rabid dog. The Silver reached for Hard's throat, bringing attention to the reason for the pressure in his neck—he had a machete buried in the side of it. They'd been trying to decapitate him when he woke up.

There was only one person in Silver City that was known to have a kink for skullfucking, and it just so happened to be the one man that may have wanted him killed if he thought a Tricks hitman was stalking their turf.

Archie 'The Mole' Tanner.

Hard drove an elbow into the Silver's skull, bursting it like an overripe melon, and dragged himself to his feet to finish off the mime. But the mime must have slipped past in the commotion as in that moment, he heard the van in the back alley wheelspin and tear away from the scene. Using the exposed bone of his arm stumps, Hard took a moment to peel the decapitated Silver's mouth away from his thigh and drop it to the ground with the rest of its cooling corpse.

Now that he knew who and what he was dealing with, a plan began to formulate in the recesses of his ruined mind.

The woman they called Pounder groaned.

Sam Pounder was having one of her many reoccurring dreams, a semi-accurate memory from her time in Iraq. This one, unlike many of the others, was pleasant. More than pleasant.

"Fuck yeah, Sammy. Scissor me harder!"

Fern had been Pounder's first sexual encounter with a member of the same sex. She'd always assumed scissoring was something people joked about but never actually did. Now she was finding out how wrong she was firsthand. After sharing a powerful orgasm, Fern crawled up to lie alongside Pounder and rested her head on her bosom. "Do you think any of this is real?"

Pounder lifted her head from the pillow and looked down at her lover with amusement. "What do you mean? Life?"

DEAD HARD

Fern chuckled. "No, silly. This," she motioned between them with her hand. "Us. Do you think we'd still be a thing if we were out in the real world?"

"I don't see why not," Pounder said, relaxing her neck and staring at the ceiling. She wished Fern would be happy with things being casual, carefree, but she always seemed to want more, and Pounder didn't know how to tell her she wasn't looking for anything serious. She opened her mouth to speak, but nothing came out.

Something wasn't right; this wasn't how it was supposed to go. Fern raised her head to look at her. Pounder tried to scream.

The flesh of her lover's face was peppered with metal shrapnel and peeled back across her brow and chin. The skin at the corners of her eyes split, ripping across to her temples as she smiled.

"What's the matter, Sammy baby? I tried to warn you before it was too late." Pounder looked to the window. The sky had turned inky black, flecked with red. "No matter. At least we'll be together again soon."

Pounder stared into the dead woman's eyes and felt herself fading away.

The temptation to crush her head beneath his knee was almost unbearable. Hard somehow managed to fight the urges for long enough to control himself.

At first, he'd attempted to hold his severed hands between his knees and mash the two stumps against them, hoping the black blood would reconnect them. After several minutes it became clear it wasn't going to work. He let them slap back to the floor and reached for the notched handle of the blade stuck in his neck with both stumps. The wound bled surprisingly little as he wrenched it free. He wondered what would happen when he ran out of blood. With any luck, he would be able to die for real. Until then . . .

Hard resumed his position on the floor once more, knees raised, the machete blade between them pointed at the floor between his feet. There was a wet sucking sound as he forced the stump of his right arm over the blade, all the way up to the hilt. The pressure inside his forearm would take a little getting used to, but it didn't hurt in the slightest.

With one arm taken care of, he scanned the room for anything that may be suitable for a makeshift hand. Nothing. Glowering, he returned to Pounder's body and forced the stump of his left arm into her drooling mouth, lifted her effortlessly and carried her back upstairs and into the door on his left. After lowering the woman to the floor, he positioned her head between the pillars of a plastic bagging machine. Activating the device with the tip of his bladed arm, he watched as the machine lowered a plastic sheet on either side of her head. The sides closed in, and vacuum sealed both her and his lower arm. He lifted her once again, then slammed her down across a steam press and tore his arm free of her mouth.

She looked like a sex doll; her panicked features exaggerated beneath the tight plastic, her mouth pinned in a wide O. Pounder made a series of muffled moans and went to bring her hands up to her face. Hard slashed at her arms each time she attempted it.

"LET ME UP! I CAN'T BREATHE!"

Hard was pinning her to the board with his arm stump and increased the pressure until she stopped speaking. Her arms were covered in red trenches leaking life over the tiled floor. After a brief moment in which he thought he'd accidentally killed her, she took a rasping breath and said: "W-what yuh wuh-want."

He'd successfully broken her.

He poked a ragged hole in the plastic over her mouth and said, "You're going to tell me where I can find Feltstone and Tanner."

Pounder wheezed, then coughed up a lumpy concoction of blood and phlegm that ran down the sheeting like oil on water. "What makes you think I'd know where that fuck . . . Feltstone is?"

"Where is Tanner."

"I can't—" Pounder screamed as Hard pressed the tip of the machete to her collarbone. "OKAY! Shit, man, what did they do to you? You ain't like no Trick I've never seen."

Hard sneered. "Still think I'm one of them, huh? Just answer the fucking question."

Pounder's eyes searched frantically beneath the tight plastic, the realization dawning on her slowly. "You're not . . . are you . . . No. I don't know who he is. A freak. Just SHUT UP AND LET ME THINK."

Hard narrowed his eyes. Was she losing her mind? Had she lost too much blood already?

"Fuck it," she said finally. "Tanner is dug in at his mansion up in Silver Hills. Big fucking gated place. You see the mole statues, you're there."

"And Feltstone?" Hard demanded, applying just enough pressure to the machete to pierce the skin.

"I ALREADY TOLD YOU, MAN! I DON'T KNOW. LET ME FUCKING GO!"

The first light of day began to creep in through the small windows at the side of the room. He didn't particularly want to be here when anyone else showed up.

"Please." Begging now.

"If you don't know where Feltstone is, tell me where I can find the Tricks."

"Eastside Labs, I guess. Same place as always." She paused, seemed to perk up a little. "You goin' after them?"

Hard intended to apply a little more pressure to the machete but was still not used to wielding the tool implanted in his arm and applied a little too much pressure, creating a two-inch hole in her larynx. Pounder lurched against Hard's stump as he withdrew the machete and studied the winking gash with interest. He held her in place as she convulsed, her screams becoming increasingly wet.

Hard reached across her with his free arm and pulled the top half of the clothes-press across as the plastic mask slowly filled with gore and obscured her features. He removed his stump from her chest, quickly snapping the clothes-press shut over her head. The room filled with the smell of melting plastic and crisping skin. Fortunately, Hard could not smell anything. He held the press in place until her body stopped fighting. Then, after another thirty seconds or so, he lifted the lid and allowed her to slide to the floor. Pounder's face was a ruined ball of stringy plastic and welted flesh.

As he walked out of the room, he could have sworn he'd heard another female voice sobbing her name. He turned around to check.

There was no one there.

Chapter Eighteen

Two lime-green vans patrolled the only road in and out of the Eastside Labs. High above, on the struts of several rusted cranes, Sonny could see the glint of the rising sun as it reflected off the sniper's rifle scopes. He tried not to pay them any attention, but it was easier said than done when he knew they were watching his every move.

"I must admit," the ugly Tricks bastard said as he led the mobster down a set of near-vertical steps running parallel to the decline of the road, "I'm a little surprised to see one of you's lot here. Especially alone."

"You're telling me," Sonny grunted. "I assure you, I come in peace. Boss just has a few questions need answering, is all." He stuttered, trying to maintain his tough-guy persona. It was proving difficult, given the circumstances.

"Well, you'd better hope Black Eye is in a good mood. He don't take too kindly to unannounced visitors."

They made it down the stairs without anyone dying and followed a wet passageway through a maze of shipping containers to a lone building at the far end. Two burly men stood on either side of the door. Both wore tinted shades that failed to hide their glares as Sonny and the others approached.

"Here to see Black Eye," Sonny's guide said to the one on the left.

"He armed?"

Sonny shrugged before pulling an Uzi submachine gun from the shoulder holster beneath his shirt. He handed it to the big guy. The man to the right of the door—a black man with a series of jagged scars across his shaved dome—stepped forward and patted Sonny down. Seemingly satisfied, he stood and delivered a lightning-fast punch to Sonny's right eye.

"What the fuck, man!" Sonny cried. He was used to dealing with pain, but the blow had taken him off guard.

"Relax," his guide said. "You can't see the boss without it."

The two guards dropped their shades just enough for Sonny to see that they, too, were sporting black eyes. Sonny squinted against the swelling, which was immediate and intense.

"You going in with him?" the black guard asked Sonny's guide while cracking his fingers.

"Nah, man. I'm good." He slapped Sonny on the back. "Good luck in there."

"Thanks. I guess."

The men stepped aside to allow entry. Sonny passed through the door, stumbling a little as he found his balance. "Assholes," he muttered, "Could'a least warned me."

Reggae music was blaring from somewhere ahead. The air was thick with a high-inducing combination of cannabis smoke and chemicals. He ambled forward, trying to avoid eye contact with the men and women in the rooms he passed. Some were cracking and bagging large sheets of crystal, others counting wads of cash. He concluded they must do the cooking in one of the other buildings dotted about the place, as he saw no equipment. As he reached the end of the corridor, he found the source of the weed smoke.

A small, nerdy-looking kid no older than fifteen sat on a beanbag surrounded by joint roaches and energy drinks. He paused the video game and placed the controller on the coffee table before him, spending an unusual amount of time getting it perfectly straight. "Come, come," he said, beckoning with one hand.

Sonny looked around. It looked like a typical teenager's bedroom, minus the bed. Posters of topless models and sports cars were plastered across every inch of the walls. Several balled tissues were strewn about the far corner next to a laptop and a bottle of hand lotion, and beneath the chronic was an underlying scent of sweat. Was he in the right place?

"Black Eye?"

"None other, homie," the kid said. He removed his bottlecap glasses and polished the lenses with the bottom of his food-stained t-shirt before replacing them. Sonny had no idea how the kid had earned the moniker. "What can I do for you? Beer? Smoke? You a gamer?" he asked, nodding to the TV, then readjusting the position of the controller on the coffee table.

"My name is Sonny. I work for the Mole." He half expected the kid to pull a gun on him at the mere mention of the name, but he didn't so much as flinch. "We'd like to know if a man, one Julio Bandera, is still under your employ."

"Julio? What's he been doing now?" The kid cracked a can of energy drink and took a long drag before continuing, "Julio is one of my best homies. You're not about to tell me he's been causing trouble for you lot, are you? That'd break my heart, man. Really. He might be a stone-cold killer, but he makes a mean omelette. Hey, did you ever hear about the time he and I went to Tijuana, Mexico? We'd had a few drinks too many, right? And there was this man on a street corner wearing a donkey suit—"

"Please," Sonny said. He had better things to be doing than listening to the ramblings of a stoned kid. If this was really the infamous Black Eye, leader of the Tricks, then why hadn't Tanner wiped them out once and for all a long time ago? "I just need to know what your man has been doing the last week or so."

Black Eye sighed dramatically and pushed himself to his feet. All five foot six of him.

"Did you just interrupt me in my own house?"

"I didn't mean no disrespect, Mr Eye. Jus' we got a real situation going on, and the boss wants me to get to the bottom of it."

The kid stared Sonny in the eye for an uncomfortable amount of time. Sonny wasn't one to back down from a confrontation, but time and place . . . he let his eyes drift to the television, and the dark-stained sledgehammer alongside it.

Black Eye laughed. "I'm just playin', man! Jeez. You should grow some balls."

Sonny somehow managed to resist the urge to lurch for the kid and crush his throat with his bare hands. "As I was sayin', Mr Black Eye, I'd just like a little information, then I'll be on my way."

"And what are you going to do for me?" the kid said, clearly enjoying himself.

"I'm sure Tanner will reward you for your cooperation."

"I don't need anything from the 'Mole,'" he said, face twisting with disgust. "And quite frankly, I'm not sure why I should entertain you any longer."

Sonny made as if to shake the kid's hand and leave, but then headbutted him as hard as he could, shattering his glasses and sending him tumbling over the back of the beanbag. He was on him before he could react, covering his mouth with the palm of his hand. "You're going to tell me what I need to know. Then, I'm going to knock you out. You'll cooperate if you ever want to wake up again."

Blood ran down the kid's face from the split across his forehead.

"Blink once if you understand."

Black Eye blinked once. Sonny withdrew his hand. There was no telling the repercussions of what he'd just done may have for their crew in the long run—but what's done was done.

"You lot are all the same," the kid said. "Always throwing your weight around as if it's the answer to all your problems."

"Listen, kid. I don't take no pleasure in hurting you, but if you'd just answered the question, I would have been gone by now. Truth is, we've got some pretty serious shit going down, and I have it on good merit that your guy is the cause."

"You don't know what you're talking about. Bandera hasn't left my sight in months. He's my personal bodyguard."

"Never there when you need them, though, huh," Sonny said. He knew the kid was telling the truth. Or what he believed to be the truth.

"I think I'm doing just fine without him here."

"You must have a pretty funny definition of fine, kid. As much as I'm enjoying myself, it's time to say goodnight." Sonny knelt on the kid's chest and drew his fist back for the knockout blow. "Ah, shit!"

The puckered skin around his bruised eye pulsed violently. For a moment, he forgot all about the kid as he touched a hand to the hot, tender flesh of his lower eyelid. The pain was searing. He blinked through tears and shot to his feet when he saw Black Eye's face.

The kid's eyes were whirling black holes. Thick tendrils of black corruption were spreading across the rest

DEAD HARD

of his face. "What's the matter?" he asked, standing and removing his broken glasses. "Didn't your boss tell you how I got my name?"

Sonny was unable to reply, the pain in his eye reaching immeasurable levels. He stumbled sidewards, squeezing the inflamed flesh with both hands in a desperate attempt to satiate it. Any worse, and he would tear it out himself.

As fast as it had begun, the pain subsided, but his head was throbbing. He took several deep breaths, then wiped the tears from his face. When he looked up, Black Eye was back on the beanbag with a fresh joint between his lips. The sledgehammer was resting across his legs.

"You understand I can't let you leave," he said.

Sonny thought he might have been able to flee the building before being tackled, but there was no way he was getting past those snipers. "A simple misunderstanding," he said, trying to maintain his composure. "I'm sure you understand."

Black Eye lit the joint, then reached for the video game controller. He blew a large cloud of smoke, unpaused the game, and began bobbing his head to the Reggae music playing through the speakers while his character drove through a city, shooting pedestrians.

Sonny shuffled awkwardly on the spot. The kid didn't seem to be interested in his presence anymore... but an imposing figure blocked the doorway as Sonny turned to leave. "You..."

Bandera showed no recognition. He had a fat, purple eye but appeared unharmed otherwise. No bullet hole in his forehead. No indication he'd recently been involved in a fiery shootout at the SCPD. It wasn't the same man.

Not that he would ever get to report this to Tanner.

He didn't even have time to beg before the .9mm round blew a hole in the back of his head and splattered his brains across a poster of a busty pin-up model.

"Damn it, Julio!" Black Eye screamed, tossing the controller to the floor. "That was my favourite poster! How many times do I have to tell you to take them outside before killing them?"

"Sorry, boss," the hitman said. "I didn't like the way he was looking at me."

Black Eye shook his head. "Get him out of here."

"Boss."

Bandera picked up the mobster's corpse and slung him over his shoulder fireman-style.

"What was that?"

"I didn't hear anything, Black Eye, sir." There a loud crash, followed by a gunshot. "He must have brought backup. They're making a move on us."

"Forget about him. Get out there and make sure you get every last one of them."

Bandera tossed Sonny's body into the hallway and ran out of the building with the rest of the Tricks to join the fighting outside. A terrified scream and several gunshots travelled down the corridor before the door closed behind them. Black Eye continued smoking as he moved to the window blinds. He wasn't overly worried—he had confidence in his men. This was a long time coming, after all. Strange way to start a turf war, sending one of theirs in unarmed, but the Mole had never been one to be predictable. Perhaps he'd been banking on that idiot taking him out.

"The hell . . ." Black Eye muttered as he parted the blinds and peered through. He couldn't see the entirety of the grounds from his office, but what he *could* see was perplexing, to say the least. The only road in was clear of vehicles, which meant the Mole's men must have entered on foot. Or from the air? Either way was a severe misplay by them. Second, and perhaps most concerning, was how no one appeared to be shooting back at his men.

He watched as one of his own shot up from behind a red shipping container and rocketed through the air toward a sniper posted above. The pair collided with enough force to knock them both off the crane and come crashing to the ground just short of his building. Their bodies were melded, twitching messes. The man on top, facing him, fixed his wide, bloodshot eyes toward the window, locking eyes with Black Eye as the life drained from him.

Several white-yellow flashes reflected off the containers as his men unloaded their assault rifles at the unseen targets. It fell silent a moment later. A stack of three shipping containers shifted sideways several meters as if a truck had just been driven into them, the top of which fell and split against the concrete with a heart-wrenching clang. Several months' worth of cocaine exploded into the air.

The fight continued out of sight, heading toward his location, and for the first time in his life, Black Eye began to panic. He could only hope his men could stop the attacker's approach before they reached him, or that Julio and the others would at least have achieved the one imperative directive that applied to every violent encounter.

Give them a black eye.

Black Eye flinched as what sounded like a rhino tore its way into the front of the building. A moment later, Julio tumbled through the door. Black Eye couldn't tell if he were alive or dead. He snatched a Beretta M9 pistol from beneath his beanbag but kept his lucky sledgehammer in his other hand. He pointed the gun at the doorway. "Julio?" he said, eyes flitting between the burnt policeman with the sneer on his face to the bodyguard face-up on the floor. "What is this? Who are you? Why do you look like Julio?"

The policeman remained silent as he stepped into the room. Black Eye grit his teeth as he realized how wounded the man was. His eyelids were swollen. That was all that mattered. As long as this policeman was the only one that'd made it through his men alive, he was in no danger. He maintained the ruse of being scared (although it *was* a little creepy that the officer was making no move to either shoot or arrest him). Black Eye kept his Beretta trained on the policeman's chest while his body tingled like a cold shower on a hot summer's day, a semi-erotic feeling that stemmed from his chest and cumulated in his pupils. He focused on the officer's eyes. Through his murk-clouded vision, he watched the policeman's eyelids bloat to the size of watermelon slices. The man showed no reaction. Now Black Eye was scared for real. He increased the intensity of his stare and felt a little flip in his stomach as the dark purple flesh around the man's right eye ruptured and spilt a black, gunky muck across the floor.

The officer ate a shot to the chest as he closed the gap between them and drove the sole of his boot into the front of Black Eye's knees. Black Eye squealed as he collapsed to the ground beside Julio, his lower legs trapped beneath him and his sledgehammer tumbling beneath the coffee table. The man kicked him onto his back. "No!" Black Eye screamed, only now noticing the blade protruding from the man's arm like something out of an 80s slasher. He raised his hands to protect himself from the steel and gasped as his fingers slapped across his face in a spray of crimson.

The burnt police officer was smiling as he used the blade to penetrate the first of the kid's eyes and pluck it from the socket. Black Eye's remaining eye rolled back in his head before it, too, was pierced and snatched from his skull.

Hard bent to peer inside the kid's orbital cavity the moment the second optic nerve snapped, then inspected the eye on the end of his machete arm, but could see nothing unusual with either. He shrugged, then brought the blade down on the middle of the kid's face.

Chapter Nineteen

Tanner paced the expanse of his bedroom aimlessly, waiting for news. It had been over twenty-four hours since he'd heard from Sonny and the other idiots, and their phones were going straight to voicemail. He stopped beneath the floor-to-ceiling windows overlooking Silver City with his hands folded behind his back. Somewhere down there, past the bristling cedars and crumbling sandstone dropoffs, was a man hell-bent on bringing down the criminal empire he'd spent his entire life building. He'd had a plan from the get-go, applying for a position in the SCPD to gain power and control over his rivals legally, before using that influence to take down the mob's leaders and slip in before the dust had even settled. Call it what you will, but Tanner had always been an enterprising man. Far from stupid, he knew that crime *does* pay, as long as you have the contacts and cash to cover yourself.

How was it possible that all of this could be picked apart by one man? A Tricks hitman, no less? He'd always known the Tricks would rear their ugly heads again after he'd organized Cue Ball's little accident, but so soon? The kid couldn't have barely hit puberty, but he had balls. That much was certain.

Tanner sighed as he moved to the marble pedestal in the middle of the room and withdrew one of his special cigars from the wooden box. Lighting it, he appraised his collection of heads, many of which had been mounted and stuffed with their mouths open and eyes wide, as if capturing the moment they'd realized their fate was sealed. He was all for diversity; women, men, and even a few kids. There was a blank space on the wall above the walk-in wardrobe between Cue Ball's ruined head (the taxidermist had done an excellent job of reconstructing his face—Tanner's erection never failed him whenever near it) and the head of his first wife (she sucked dick better dead anyway), where Bandera would be placed. Tanner wasn't without a sense of humour—he chuckled at the thought of Bandera's head mounted next to his former boss.

"Sir?" The henchman averted his gaze from his boss's tented boxer shorts as he spun to face him. Tanner had been so lost in thought that he hadn't noticed anyone knocking on his door. He tied his dressing gown loosely beneath his waistline and tucked his dick beneath it.

"Sir, we have a man outside the gates claiming to be Graham Krik. Says he needs to talk to you urgently."

"What do you mean, claiming to be? Are you not capable of facial recognition?"

"Boss, with all respect . . . I really think you'd better see this for yourself."

"God damnit."

Tanner tossed the cigar over his shoulder. It never hit the floor.

"What the fuck is this?" Tanner said as the henchman swung the monitor toward him.

The other man shrugged as if to say, *See what I mean?*

"Hello? What do you want?"

The figure on the monitor came closer to the camera. "Boss? That you? It's me, Krik. Let me in!"

Tanner moved closer to the screen to ensure his eyes weren't deceiving him. It was either some next-level SFX makeup, or Krik had finally lost the plot.

"What's wrong with your face, boy?"

Krik scanned the vacant road on either side of him. "He killed the Silvers! And Pounder!"

"Bandera? You expect me to believe one man took out a team of Silvers, as well as one of my best shooters?"

"Just let me the hell in!"

Tanner rapped his knuckles on the oak desk, then started to walk out of the room. "Let the idiot in," he said over his shoulder. The henchman passed a message over the radio to advise the others a friendly would be coming through, then buzzed Krik through the main gates.

"What in the fuck have you done to yourself?" Tanner asked, disgusted, as Krik tumbled through the towering front doors and collapsed onto the floor.

"It ain't Bandera," was the first thing to fall out of the mobster's mouth as he peeled the raw flesh of his skinned face away from the tiled floor. "The guy. The killer. It's not the Tricks at all. It's never been the Tricks. It looks like Bandera, but it ain't Bandera."

"Slow down and tell me what happened."

Krik gave his boss the play-by-play of the events that had transpired. His face was red-raw and slick with dirt and grit, downright disgusting and making it hard to focus on anything he was saying.

"So what happened to your face?" Tanner asked once Krik was done.

"What's wrong with my face? Can you still see the scars?"

It was then that Tanner noticed the rolls of bunched skin beneath the man's bloodied fingernails. The things he'd witnessed had been enough to push him over the edge into insanity once and for all. He felt something foreign lurking in his lower colon. Was it pity for the man? Would it be kinder to put him out of his misery?

"I want you out there with the others," he said. "Tell them what you just told me."

Krik rubbed his hands together anxiously. "I was kinda hoping you might let me sit this one out, you know, after everything . . ." he lowered his head.

"Don't make me repeat myself."

The Mole watched the last of his personal guard slip back outside and cursed. What did he do to deserve to have such imbecilic fools working for him? It was no use getting hung up on it. If what the man told him were true—his testimony was unreliable, at best—they were dealing with a new threat altogether. But if it wasn't the Tricks, and it wasn't the SCPD, then who did that leave? One of his own?

He would have recognized the fool. He *did* recognise the fool. He had been the spitting image of Julio Bandera. No. This was the work of an out-of-towner. It had to be. Perhaps he were a relation of the Trick hitman, trying to make a name for himself. Yeah. That had to be it. He took his phone from the pocket of his dressing-gown and tried Sonny's number again—it seemed only fair to warn him—but was met with his voicemail.

Whatever. Whatever was going to happen next was going to happen soon. He could feel it in the air like the rising heat before a lightning storm or the moment a crack-baby spots an uncapped needle in the street, and Sonny wasn't going to be near enough to be of any assistance, unless he was already on his way back.

Tanner returned to the security room and told the guard to put everyone on high alert. He decided to leave out the specifics of the madman's ramblings as it would only put his remaining staff more on edge than they already were.

"So it was never Bandera?" the henchman asked, looking up from his swivel chair. "Then who have we been going after this whole time?"

Tanner shook his head. "That's the million-dollar question. Hang on a second, what's that?"

"What?"

"There." He pointed to one of the exterior cameras that faced down the road toward the city. "Zoom in." The henchman pulled a keyboard toward him and tapped a few buttons to home in on the dark shape in the distance. "Son of a bitch. He's here. Get eyes on that fucker right now. I want him stone cold before he gets anywhere near those gates."

The henchman relayed the order over the radio. On one of the smaller screens, several suited figures trotted down the pebbledash driveway toward the gates. Tanner couldn't quite understand what he was seeing on the main screen—the man appeared to have some sort of long blade where his right hand should have been, and in place of his left was the shaft and end of a large hammer. The closer he came, the more uneasy Tanner felt.

First off, the man was wearing an SCPD uniform, burnt in multiple places, complete with a shattered and melted riot helmet. But melted plastic was far from the worst thing that had happened to the man's face since they'd last had a run-in; one of his eyes had been eradicated, like a ruptured water balloon, purple, loose, and wet. The rest of his face hadn't fared much better: plastered in black and red, it looked as if he'd been in a fight with a man-sized cheese grater. A dark hole on the side of his neck looked wide enough to admit a fist, yet somehow was not bleeding, and his entire body was peppered with bullet holes.

"What the hell are you?" Tanner whispered.

Hard could see the place the moment he'd set foot on the winding road leading up to the Silver City hills. Obnoxious and overly extravagant, the contemporary structure jutted into the sky like a cancerous sore on mother earth, rooted too deep to be removed.

Deciding to ditch the road to avoid the line of sight, he cut a path straight up the crumbling rockface and through the trees while stewing in all-encompassing anger, wondering if this was all that was left for him now that he couldn't ever return to his regular life. Perhaps his sacrifice was what was required for the betterment of the city. Perhaps it was the city itself that was keeping him alive and would only let him rest once he'd purged her. Lord knew it was overdue. He didn't know if it were even *possible* for him to die at this point—his body had already taken far more punishment than he'd ever imagined feasible—but only one man would be walking off this hill.

Hard stepped out of the tree line and onto the road once the regal main gates of the estate were in view, each post capped with a concrete mole. He was well aware of the security cameras tracking his every move as he drew closer. It didn't matter. Let them prepare. Let them panic. It wouldn't make a difference whether they saw him before he ended them. The gates opened inward smoothly on their metal tracks, and several men and women charged onto the street. No warning shots—they opened fire like their ammunition was infectious and they were desperately trying to be rid of it. Dust clouds swirled into the air as bullets cracked into the concrete all around Hard. A handful connected with their intended target, but the guards didn't seem concerned with opting for a spray and pray approach. Four of the five suits retreated upon witnessing Hard shrug off their assault. The final realised a little too late that his buddies had abandoned him. His cry for a spare magazine went unheard, and he was left with his pants around his ankles in the path of a bull. Hard closed in on the man and ate a fist to the face. He responded in kind, with a right hook from his bladed arm that separated the goon's face from his skull.

The wrought-iron gates clicked shut. Locked. Hard kicked through them on the second attempt and started up the paved driveway on the trail of the fleeing guards. He'd been expecting heavy resistance, but if this was the best Tanner had to offer, the mole was as good as dead.

A resounding gunshot echoed throughout the hills like a thunderclap in an auditorium, and one of the retreating guards spun off to the side and fell face-first into the base of a water fountain. Another shot, and a second guard's head was blown off her shoulders, her body back-flipped before crashing to the grass. The third and fourth guards threw their hands up and shouted something Hard couldn't make out before turning back around to face him. There was a slight movement on the flat roof of the building, and there, between a rounded chimney and a static telescope, was Tanner, a fifty-calibre Barrett sniper rifle on a tripod in front of him, just beyond the lip of the roof.

The nearest guard approached Hard with his hands held low in front of him. "Please. I-I don't want to fight."

Hard sneered. "You don't have to."

Spearing him through the chest, he lifted him off his feet and hurled him at Tanner. Tanner managed to get a single shot off before the man collided with the tripod, obliterating the man's groin and giving the rooftop a fresh coat of crimson. Upon witnessing this, the final guard broke into a sprint, but was unable to outrun the Mole's rifle.

"LAST CHANCE, ASSHOLE," Tanner bellowed. He held the sniper in both hands, the stock braced against his shoulder, and the smoking barrel pointed at Hard. "LEAVE NOW, AND YOU HAVE MY WORD THAT I WON'T COME AFTER YOU."

"Doesn't look like you're in a position to be making demands," Hard said, spreading his arms at the corpse field around him.

The man's dressing gown billowed in the wind, exposing his bloated midriff and stained white boxers. His pierced nipples pointed toward the floor, two golden, cigar-shaped bars gleamed in the sunlight.

Hard was spun to the side and knocked to the ground as he moved for the front door, with a pressure in his shoulder that was the closest thing to pain he would likely ever feel. Regardless, he was back on his feet and kicking down the doors before Tanner had even chambered the next round. He found himself in a foyer that screamed of compensation—a towering statue of Tanner among marble and mahogany and gold-framed paintings. Doors branched off in all directions, and a wide, imperial staircase led up to the first floor. He made for the stairs and was halfway up when the fist-sized hole in his shoulder began making a squelching noise. The cavity was filling with blood, or graphene—whatever the hell it was—and rapidly solidifying. By the time he'd reached the landing on the upper floor, he could no longer put his fingers through it without hitting spongy resistance. He didn't care to give it any thought. The only thing that interested him was finding a way to the rooftop and putting the Mole six feet under.

Hard walked to the end of the carpeted hallway and kicked down the door. Bathroom. The next was a large lounge with leather sofas, a wall-mounted flatscreen TV, and a man with a skinned face.

"You!" Skinned Face squealed. His eyes fell upon the wrecking tools protruding from Hard's arm. "How are you still alive!?"

Hard was too transfixed by the man's grotesque appearance to pick up on what he'd said. He had no idea who he was, but he was one of *them*. Skinned Face's eyes were cotton white in contrast to his shiny, slick face. It would have been a mercy to put him out of his tortured existence.

"W-where are y-you going?" he stammered as Hard backed out of the room.

He searched the rest of the floor with rising impatience. The final room at the far end of the hall lay behind a set of double doors with golden cigar-shaped handles. He kicked the doors open and was greeted by dozens of severed heads mounted on the walls and ceiling, like a hunting cabin if the roles between animal and human were reversed. A four-poster bed was home to even more. On the pillows, and beneath the sheets. A couple were familiar to Hard, others had seen better days and were barely recognizable as human. A walk-in wardrobe was set into the wall opposite the bed, doors open, and the bottom of a pull-down ladder was visible.

"Dad?" he mumbled, stumbling to the mounted head on the floor inside the wardrobe. It took a few moments for the relevant dots to connect—he wasn't used to seeing him with his jaw locked open and his forehead plastered in semen—but it was undoubtedly him. "Dad. . ." Hard dropped to his knees and cradled the head, tearing a shirt off the rack to clean its face.

His father's head had never been recovered. From the trauma to the surrounding tissue of his shoulders, the SCPD had concluded that it had been destroyed in the accident. It was a disgrace to know where it had really been all these years, how many times Tanner had . . . Hard set his father's mounted head down gently. Spurred on by a fresh wave of anger, he mounted the ladder.

"That's close enough," the Mole said as Hard slammed the roof hatch behind him. To his credit, Tanner allowed him to get to his feet before blowing another hole in him.

Hard paid minimal attention to the little man with the big gun. Instead, he took in the view of the city that few honest men would ever get to enjoy. From these heights, even the black smoke from the smouldering tire fires drifting throughout the broken-tooth buildings could have been torn straight out of an oil painting. Like many people in his life, Silver City seemed nice enough from a distance, until you got to know it a little better.

"Look at me," Tanner said.

Hard closed his eyes and listened to the sounds of the city; the blare of a scooter horn, the scream of a newborn crack-baby, the rumble of a building's rotten foundations finally giving way. Though he'd been through so much over the last seventy-two hours, there was still so much left to do.

"When you're ready."

Hard opened his eyes and turned to face him, his weaponized arms relaxed at his sides. "Was it worth it?"

Tanner's eyes softened momentarily. Not a word passed between them.

Finally, Hard continued, "The state of the city, since you and your men started throwing your weight around. Was it worth it?"

"It was going to shit when the mines closed anyway. I'd do it again."

Hard cast his eye over the hills once more, jaw tightening. He wasn't expecting remorse from the man, but he was unwilling to admit his wrongdoings even when faced with death. Politicians had written the city off as a lost cause the moment Tanner took control of the mob; they'd fled to places that still had law. Instead of the mob using their newfound status to return the city to at least *some* semblance of normality following this, they'd been happy to instead bleed it dry and flood it with drugs.

Tanner's phone rang. He held a finger up to Hard, then pulled it out of the front pocket of his dressing gown and answered. "Yes. Uh-huh. Oh. Really? Interesting. Thank you." He hung up. "I know who you are," Tanner said. Then, sensing Hard's surprise, "What, you didn't think I'd find out? *Dead* man? What I don't get is *what* you are, and I don't think you do either. Poor little dead Hard, blazing a course of destruction in the name of vengeance, while failing to grasp the delicate intricacies of running an operation of such a scale and how his actions may affect the rest of the city."

"Dead Hard," Hard huffed. "Whatever I am, it's thanks to you. I was quite content living my quiet life before you ordered my murder." The bloodied machete glinted as he brought the blade up to waist level.

The Mole put one foot behind the other and braced himself against the butt of the sniper rifle. "I'll blow your fuckin' head right off your shoulders if you move another muscle."

Hard remained still. Sure, his body was capable of some truly remarkable feats, but dodging or surviving a fifty-calibre round at this range? All he needed was a small opening.

"If you're so confident you could kill me, why haven't you already?"

Tanner looked up from the scope but kept his sweaty little finger glued to the trigger. "I have a proposition that I believe would be mutually beneficial."

"Not this old spiel. I'd rather die again than do anything that may benefit you."

"You may be willing to sacrifice yourself for your little vendetta, but your name wasn't the only one that turned up. Does Khoai Lang, mean anything to you?"

Hard stared down the man that was almost half his height. "You dare touch her..."

"My people are bringing her here as we speak. I'll admit, I was a little unprepared for you to turn up as fast as you did. If I'd known you were coming so soon, I wouldn't have had most of my crew out in the city looking for information on you... I'll call 'em off once I know you're on board. But know this: She will never be out of our sight as long as she lives. A modicum of bad behaviour from yourself, and I won't hesitate to have her slaughtered." The Mole shifted his grip on the weapon.

Hard softened his posture. "What do you want me to do?"

The corners of Tanner's mouth twitched, but he managed to maintain his poker face.

"All I need you to do, for now, is finish what you started, and I'll make sure you're adequately compensated. I need Chief Feltstone out of the picture."

"And here I was, thinking you were friends."

Tanner chuckled, shook his head. "The man's a fuckin' fool. I mean, he had the entire SCPD at his disposal and still managed to let one man force him into hiding."

"Not sure you're doing too much better yourself."

"Watch your fuckin' mouth, asshole. While you're workin' for me, you'll do well to show a little respect."

"My apologies," Hard said with as much conviction as he could muster.

Tanner relaxed his trigger finger. "That's better. Now, word has it—"

Hard lurched forward with explosive speed and buried the machete in the Mole's shoulder. He'd been aiming for the skull, but the little man jerked just enough to avoid the death blow and squeeze his trigger. Hard roared in surprise more than anything else as a football-sized hole blew out the back of his chest and tossed him to the ground. Tanner fell on top of him, the blade still lodged in his clavicle, and screeched like a wounded cat as he tried desperately to free himself. Paying little attention to his thrashing, Hard gazed up at the contrails crisscrossing the cloudless sky, a gentle breeze licking at the hole where his heart had been. Would he ever get a chance to find out what was up there? What if the bullet had hit him in the skull? Would he have continued without a head? Without a brain?

After a minute or so, he rolled the Mole off of him, tearing the blade free of his shoulder in the process, and pulled himself to his feet. Tanner attempted to drag himself along the pebbled rooftop to where the rifle had fallen, but Hard stamped on his ankle, obliterating it and pinning him in place.

"FUUUUUCCCCK," Tanner screamed as he rolled over. His ankle gave off a sound akin to popping twigs in a campfire as it turned one-eighty. Frothy spittle sprayed across his red-mottled cheeks. "YOOOUUUU!"

Hard twisted the shattered bone beneath his boot. "Where is she?"

"FUUCCCKK Y—AAH!" Tanner screamed as Hard stabbed him through the abdomen and wrenched the blade down to just above his groin.

He removed his foot from his ankle and lowered himself until they were eye to eye. "I'm not going to ask again."

"GET UP HERE! HE'S KILLIN' ME!"

It took Hard a moment to notice the earpiece, the very same moment it took for the roof hatch to slam behind him and his fiancée to appear in front of a skinny man in a tight-fitted waistcoat.

"Don't try anything," Hard said, glancing over his shoulder.

The Mole groaned, twitched ever-so-slightly on the length of metal.

Something inside Hard was telling him to kill him regardless of the consequences, that if Lang were to die, it would be worth it. But the smaller voice, the one he believed was his own, was telling him he would never, *should* never consider that as a possibility.

"Get this fuckin' thing out of me!" Tanner screamed.

Hard twisted the blade slightly before wrenching it free of the man's bloated gut. Judging from the amount of gore already pissing from the jagged wound, he would bleed out within the hour if he didn't receive urgent medical attention—even with both hands mashed against the injury, blood was oozing through. Hard turned to face Lang and got a good look at her captor for the first time. His skin was pale, his face gaunt. High cheekbones and sharp chin. Dark sockets housed even darker eyes, which twitched erratically, seemingly unable to stay still. His mouth was slightly parted, cracked lips framing yellowed, uneven teeth.

He was an adult crack-baby. High functioning. Rare, but if anyone were to keep one under control, it was the Mole—with a limitless supply of crack, the man would likely do anything to protect his boss.

The adult crack-baby kept Lang in front of him, both barrels of a sawn-off shotgun pressed against the back of her head. She was gagged, and a crescent of blood was drying beneath her left eye.

Hard stared at the man. Living so close to a crack-baby lair, he'd had first-hand experience with how unpredictable they could be. The adults he'd seen tended to be volatile and suffer from terrible hand tremors and body spasms, but he'd never seen a high functioning one in the flesh.

Beneath the shrink-wrap-tight skin of the man's face, nerves rose and fell like the bars of a digital soundwave. It was almost as if the crack itself was alive inside him, looking for somewhere thin enough to breach.

"Let her go," Hard demanded. "She has nothing to do with this."

Tanner moaned in agony.

The adult crack-baby looked past Hard. "Will you live?" His voice was distant, like sand through a metal chute.

"Not if you don't deal with this situation RIGHT FUCKIN' NOW."

The crack-baby cocked the hammer of the sawn-off.

Lang closed her eyes, a dark stain appearing across the front of her jeans.

"Don't." Hard readied himself, the scene before him playing out in slow motion.

Like a koala through a jet engine, Lang's head was fucking obliterated.

Her forehead swelled, pushing her eyes out and off to the side. A thin red line then appeared from the bridge of her nose to the top of her forehead before her cranium ruptured entirely and time sped up, the whole hot mess smacked Hard across the face and upper chest. He roared, lunging for the crack-baby with his blade extended.

The crack-baby dodged the attack with ease, appearing somewhere off to Hard's left and causing the machete to instead gouge a trench across the chest of Lang's falling corpse. Tanner managed a weak laugh as Hard spun his hammer arm and parted the air where the crack-baby had been standing less than a second before.

He fell to his knees.

Something wasn't right.

Looking down, he saw a flash of silver and a wide gash appearing in the fabric at the front of his thigh. The skin beneath was torn, but bled little. Another flash and another cut latticed over the first, gouging a large chunk of flesh. Hard swiped the machete at nothing as the crack-baby continued to run circles around him, slashing and stabbing manically. Eventually, Hard landed a lucky swing with the sledgehammer and took the man's legs out from under him with a sickening crunch. He attempted to stand, but his shins had been shattered, and he fell straight back to the roof. The six-inch hunting knife clattered away from him.

Hard got to his feet. His left leg was in ribbons, but he could still walk with a little caution. Black blood poured from his forehead, obscuring his vision as he approached the skinny freak, which had managed to claw its way over to its boss.

Tanner's breaths came in strained hitches. His hands had dropped from his sodden abdomen and lay at his side, slick with his essence. Although his eyes were dull, they still tracked Hard as he wound up the sledgehammer, then brought the business end down on the crack-baby's head with his full force. His skull exploded like a lanced boil. Hard wrenched the hammer out of the hole in the roof and turned his attention to the Mole.

"I cngivyumoneh," he sputtered between mouthfuls of blood.

"Do I look like I need money?" Hard said as he worked the machete beneath Tanner's knees and the arm of the sledgehammer beneath his neck. He lifted him, ignoring his weak cries, and walked him to the roof's edge at the rear of the property. Without hesitation, he tossed him toward the rocky drop-off. Tanner appeared to find a

MATTHEW A. CLARKE

sudden reserve of energy as he fell, arms and legs flailing wildly in all directions as if it would increase his chances of a safe landing.

It didn't.

Chapter Twenty

Detective Schafter's sick leave was technically over, but there was no office to go back to. Chief Feltstone had been in contact over email, dishing out jobs to any officers that were available/still alive, but for the most part, he hadn't had contact with anyone else. As payroll had gone up in the fire, along with everything else, it was yet to be seen if and when his wages would hit his bank account, and it wouldn't be long before he'd burnt through his meagre savings.

He sat in his kitchen/diner/front room with a bad taste in his mouth that even the cheap whiskey couldn't remedy. Something had been bothering him for a while now, and even in his inebriated state—something that he thought would help him come to a decision one way or another—he found he still couldn't choose. Hard had taken out one of the worst people Silver City had ever had the misfortune of spawning, but with Lang dead, even that felt like a hollow victory.

She'd been an innocent in all of this. Sure, she'd known the dangers of going after Hard on her own, and he'd tried to warn her off, but it didn't change the niggling feeling that he was partly to blame. He should have stopped her. Even if it had been against her will, she would still be alive today. As for Hard . . . well, there was no saving Hard. He was responsible for multiple murders, despite his intention. Something else Schafter may have been able to prevent had he had the courage to pull the trigger back in the aquarium. Then again, it was clear from their first encounter that the man should already have been dead. The couple's story was just another sad score in the city's ever-growing tally.

But it wasn't too late to see things through.

Schafter had made many questionable choices for a detective over the past few days. Many of which either directly or inadvertently led to illegal acts. Deep down, he knew it was a necessary evil, that there was not a glimmer of hope for the city without someone like Hard to do what needed to be done, but right now, tensions were higher than ever as everyone waited to see if and when someone would step into Tanner's shoes and crown themselves the new drug baron of Silver City.

Fuck it.

Schafter picked up the phone and dialled before he had a chance to doubt himself any longer. After three rings, a gruff voice answered.

"Feltstone. What you got for me, Schafter?"

"I know where he is."

Chapter Twenty-One

Schafter's nerves were at an all-time high as he heard the rattling of the approaching engine. He watched the marked sedan in his rear-view mirror as it bumped up the curb and stopped just short of his bumper. A plain-clothed figure stepped out and quietly shut the door behind them. Schafter took a steadying breath before stepping out to join him.

"You're sure about this?" Feltstone asked, his upper lip curling in disgust as he appraised the decrepit aquarium.

"I'm sure, sir. He's in there."

Feltstone lowered his eyes to the detective and placed his hands on his hips.

"You know, I wasn't sure if I could trust you, Schafter. But you're doing the right thing. Benjamin Hard must be brought to justice for what he's done." He went to the boot of the cruiser and grabbed a pump-action shotgun, cocked it with one hand. "Let's see that fucker keep going when I hit him with these taser rounds. I take it you're adequately armed?"

Schafter popped the strap on the holster on his belt and tapped the grip of his revolver.

Feltstone raised an eyebrow. "You've seen this guy in action. That's not going to do shit."

"I figured there would be more of us . . ."

"We're spread too thin as it is. The only guys with the balls to answer the damn radio are across the other side of the city dealing with an active shooter."

"You want to wait for them?"

"And risk him slipping away again? Fuck that. This asshole is mine."

Schafter knew the Chief would never admit the real reason for his impatience—he wanted to get Hard before Hard got him. He acted as if he didn't already know the front door would be sealed tight as their boots crunched across the bloated tarmac of the car park and ascended the stairs.

The Chief turned to him, "I thought you said you followed him here? This is rusted shut."

"I did."

"Well, how the hell did he get inside?"

"There's another door, this way," he nodded, ignoring Feltstone's impatient mutterings as he stepped over the fallen octopus arm on his way to the staff entrance. The chains that secured the doorway until Schafter's last visit now snaked across the ground like a metallic parasite. Feltstone nodded Schafter to take point, then readied his shotgun.

The door, hanging on by one hinge, caught on the uneven tiling on the other side as Schafter nudged it open with his shoe and stepped into what had once been a staff room. A clear path appeared to have been cut through several overturned chairs and tables to the door at the far end. Schafter paused to draw the Chief's attention to a faint line of black sludge. Feltstone nodded without taking his eyes from the doorframe ahead. They followed the sporadic trail down a wide hall lined with dust-covered 'Employee of the Month' frames and through a dressing room of sorts. Schafter rounded a row of lockers and almost had a heart attack when he came face-to-face with a sagging Samuel Starfish costume on a wooden bust. Several others behind held various other cartoon characters.

"Keep your shit together," Feltstone whispered.

Schafter swallowed hard and kept moving, careful not to knock any of the busts.

They followed a cold concrete passageway and found themselves standing between rows of raked seating in a small auditorium. A skylight dome illuminated drag marks in the dust on the floor, faded banners on the walls, and glints of broken bottles. And dead ahead, sat just inside the reach of the lapping shadows—

DEAD HARD

Feltstone touched Schafter gently on the shoulder and moved past, shotgun raised. Hard was facing them, but his eyes were closed. Wearing the tattered SCPD uniform and shattered riot helmet, the only sign of life was his shoulders moving in synch with his breaths.

The police Chief lined the barrel of his weapon up with the sleeping man's face and fired.

Hard face-planted the floor, grunting and convulsing violently from the taser round embedded in his cheek. Feltstone pumped the shotgun as he closed the gap between them. He fired again, this time hitting the man between the shoulders. Schafter stayed back while Feltstone tossed the shotgun aside and pulled a Desert Eagle magnum from the waist of his trousers. He approached Hard, kicked him violently in the head, enough for him to roll onto his back, and put three into his chest in quick succession, followed by one in the head. Only then did Schafter begin to approach.

"Wait," Feltstone said, peering into the bleeding face of the corpse, "this isn't the guy that attacked the station. What the fuck is this, Schafter?" He spun to face the detective, his forehead rumpled, eyes swirling with confusion.

"Sorry, sir. Things won't start to get better if you're still in charge."

"You sonofa—" Feltstone raised the magnum but seemed to sense the sledgehammer whizzing toward the back of his head and threw himself toward the shotgun on the floor. The battered visage that was Benjamin Hard stepped out from the shadows and exploded Feltstone's thigh with a downward strike from his sledgehammer arm. The Chief screamed but managed to snatch the shotgun and roll over. Hard brought both weapons down at him, but was hit in the throat with a taser round and went rigid. He dropped sideways to the floor.

Pump. BANG. Pump. BANG. Pump. BANG. Pump. Click.

Out of ammo, the Chief dropped the shotgun and dragged himself over to Hard's twitching frame. It reeked of burning flesh and ... something worse. Something chemical and sour. Feltstone, bleeding all over the floor, buried the Desert Eagle in Hard's smoking temple. "Let's see you come back from this one."

MATTHEW A. CLARKE

A single shot was all it took to kill him.

The blast reverberated through Schafter's chest and quickened his pulse like a syringe of adrenaline to the heart. He holstered the revolver with a (somewhat surprisingly) steady hand and addressed the steaming hulk, "Hard?" No response. Schafter failed to avoid glancing at the exit wound between the Chief's eyes as he stepped over him. "BENJAMIN."

Chest compressions would be utterly useless, as would feeling for a pulse. Instead, he slapped the dead man with an open palm across the face. Again. Hard's hand shot up and wrapped around Schafter's wrist on the wind up for the third slap. "I thought you were dead."

"I am," Hard smirked.

Chapter Twenty-Two

One Year Later
Felix Schafter killed the rumbling engine, kicked open the door, and climbed down the wheel of the restored Bigfoot monster truck. He'd picked it up from the garage earlier that morning and had been blown away by the aquatic-themed decals—the octopus tentacles that wound around the windscreen and the shark bite on the driver's door especially. Silver City was far from saved, but with the money and resources the Mole and Feltstone had amassed over the years now in his possession, it was getting there. The monster truck had been his little bonus to himself for a job well done.

He entered the Schafter's Private Defense building—built on the plot that had once been home to the SCPD—and scanned his badge at the security panel alongside the internal doors.

"Morning, Krik. How're things looking out there?"

Krik looked up from the computer monitor and offered a toothy smile from the mass of scar tissue that was his face. "Not too bad, sir. Robbery at the ATM on Dane Street and a bar fight on Hale in the early hours of this morning. One hostile death, three injured and in holding."

Schafter returned the smile and asked the receptionist to let him know when his nine a.m. arrived, before going through to the shared office. Despite doing what he could for the guy, he still felt terrible for him. The ex-mobster spent the majority of his life suffering from self-doubt because of his facial scars, which had led to him falling in with the wrong crowd. Schafter had found him back at the Mole's mansion and taken him to the emergency department. They'd done what they could... but it wasn't pretty. Regardless, he'd accepted him for who he was, and it turned out, that's all the man had ever wanted. Krik had been a model employee and generally good human ever since— besides, if Schafter locked up everyone that had been involved in the criminal underground, there would be no one left. He was happy to accommodate those willing to change but would come down hard on them if they betrayed his gesture. Literally.

"Good work last night," Schafter said to the man behind the desk across from him. His bulky mass looked comical on the small swivel chair, though he looked a million times better in the custom-made suit than in the police uniform.

Dead Hard shrugged, smirked. "Give me a *real* challenge."

"With any luck, there won't *be* any more real challenges."

Hard nodded before turning his attention to the picture frame to the left of his desk. He'd felt empty since his fiancée was murdered in front of him, and no amount of 'challenge' would ever be enough to distract his mind from that. Schafter had doubts as to whether the old Hard would have been able to understand that. He'd been lucky to bring it under control before the city itself turned against him and had even had the sense to replace the machete and sledgehammer embedded in his arms with prosthetics.

Schafter powered up his tablet and flicked through several of the city-wide camera feeds. From there, he could zoom out to a birds-eye view and would be able to see any crime that was reported, marked on the map by Krik, in real-time. All was quiet.

"You should probably get ready, huh? Big day for you," he said, pulling the big man's attention from his stupor.

"I guess you're right."

He left the room and returned a few minutes later wearing a sumo mawashi.

Hard had been dominating in several combat and strength sports in the months prior. At first, there had been speculation as to whether a dead man should be allowed to compete, but after much debate, officials decided that as he was still technically human, there were no rules against it. Several athletes had attempted to replicate Hard's death in the following weeks and months. None succeeded.

Schafter chuckled. Despite his ruined appearance, seeing his friend in a giant diaper never got old. "Break a leg."

Hard couldn't help but smile. "I will. But it won't be mine."

Afterword

Dead Hard is my fifth bizarro book, and for this one I wanted to try and pay homage to some of my favourite movies growing up. Michael Myers, Jason, The Terminator, Robocop. All of these served as the inspiration behind this tale, and encouraged me to be a little more descriptive in some of the gorier scenes.

I started out with the idea of someone being murdered in a case of mistaken identity and coming back as an unstoppable killing machine to wreak chaos and claim vengeance upon those responsible. I usually like to have a rough outline of where I'm going with a story when I sit down to write, but that was about as far as the initial idea went with this one. I hope it doesn't show too much.

I have several projects in the pipeline at the moment, so if you enjoy my work, keep your eyes peeled!

And as always, thank you for reading.

Matthew A .Clarke 30/5/22

Acknowledgments

Massive thanks to my beta readers, Heather Moffit, Melissa Potter, Marcie Robinson, Corrina Morse, and Sonia Clarke, for the suggestions and edits. You guys are the best.

About the Author

Matthew A. Clarke writes horror, bizarro, and anything in between. He has authored and published 6 novels and 2 novellas. He has also had many short stories published, ranging from humorous to the horrific.

Bibliography
Coffin Dodgers, Self-Published, 2021
Things Were Easier Before You Became a Giant F*cking Mantis, Self-Published, 2021
The World Has Gone to Turd and the Only Way to Save It Is with a Big Ol' Battle Royale, Self-published, 2021
Beyond Human, Black Hare Press, 2021

Those That Remain, Black Hare Press, 2021
Sons of Sorrow, Planet Bizarro, 2022
Dead Hard, Planet Bizarro, 2022

Connect
Amazon: https://www.amazon.co.uk/~/e/B082WJJPQ3
Facebook: www.facebook.com/matthewaclarkeauthor

Also Available from Planet Bizarro

Peculiar Monstrosities

A Bizarre Horror Anthology

A stripper's boyfriend bites off more than he can chew during a hiking trip.
A man looking for love marries a jukebox.
A popular children's character is brought to life, but something isn't quite right.
A shady exchange on a Kaiju cruise leads to catastrophic complications.

Peculiar Monstrosities is packed with fourteen exquisitely crafted stories from new and established authors of Bizarro fiction.

Featuring tales by: Kevin J. Kennedy, Zoltan Komor, Shelly Lyons, Tim Anderson, Tim O'Neal, Gregory L. Norris, Joshua Chaplinsky, Stanley B. Webb, Jackk N. Killington, Kristen Callender, Michael Pollentine, Tony Rauch, Mark Cowling, and Alistair Rey.

DEAD HARD

Sons of Sorrow
by Matthew A. Clarke
SOME THINGS ARE BETTER LEFT ALONE

Henk has been living a relatively carefree life in the city since fleeing the horrors of the town of Sorrow with his brother, Dave. Never would he have dreamt of returning.
Not even for her.
But time and banality have a funny way of eroding the memory of even the worst experiences, bringing only the better times to the forefront of recall, so when he receives a wedding invitation from the third part of their old monster-fighting trio, he finds himself unable to turn it down.
Sorrow has changed drastically from the place it once was, with the murders and suicides that once plagued the town being used as a selling point by wealthy investors to turn it into a morbid attraction for dark tourists.
Beneath the costumed mascots and smiling families, is all really as it seems? Or by returning, have Henk and Dave inadvertently awoken an ancient evil far deadlier than anything they've faced before?
Sons of Sorrow is the latest bizarre horror from the mind of Matthew A. Clarke.

Porn Land
By Kevin Shamel
OH, NO, PORN IS ILLEGAL!
That's right. Porn stars are criminals, pornographic websites are being systematically destroyed, and not even softcore or selfies are okay. And that's just in our world. It's literally destroying the magick city of sexual expression—PORN LAND!
Phil and Zed, arriving through magickal means and ill-equipped for adventure, must travel through the erotic metropolis and gather pieces of THE PORNOMICRON—a sexual spell-book that bridges our worlds. And it won't be easy. They'll have to get past a giant geisha and her samurai army, a determined detective who's after their asses, a badass dominatrix and her gang, a bunch more sexy people, a bunch of unsexy people... And even more things that will freak you out and make you horny—like a sperm monster and ambulance sex.

MATTHEW A. CLARKE

Will Phil and Zed put the book together, save Porn Land and their new friends, *and* make pornography legal in our world again? (Yes. It'd be a stupid story if they didn't. But it's *how* they do it that you'll want to read about.) It's a story about sucking, *and* not sucking. It's got hardcore sex *and* a hardcore message. It's ridiculous *and* you'll wanna rub one out to it. It's freakin' PORN LAND, BABY!

Weird Fauna of the Multiverse *A trio of novellas by Leo X. Robertson*

— A gimp becomes mesmerized by the koala at a zoo on Venus. She draws him into the battle between the purebred animal supremacy of the park's hippo owner and the anti-establishmentarian koala uprising.

— In a godless future, a rich Martian traveler hunts the former Vatican—now a hotspot for sex tourism—for his deceased wife. When he discovers a dead priest in the streets, he begins to investigate the weird plot of the city's head cyberpope.

— Supercats spend their days responding to rescue calls across their city. Since there aren't enough rescues to go around, one supercat decides to do something drastic and devious to resolve this crisis, changing the industry forever.

The stories of *Weird Fauna of the Multiverse* explore what happens to love and work when pushed beyond the boundaries of human decency.

Stories to Make You Puke Your Pants
A collection of bizarre tales by G. Arthur Brown

"Whether he's scribbling on napkins, writing online, or penning fiction, G. Arthur Brown is interested in taking the world we think we know, cracking it open, slathering it with weirdness, and twisting it into odd shapes--which, surprisingly, resemble the world more accurately than the world we wish we had. Brown's a prime example of how the weird and the bizarre can provide an active and irreverent critique of the real. This is fiction that's fun to read and yet deeply resonant." -- Brian Evenson, author of *The Glassy, Burning Floor of Hell* and *Last Days*

Twelve stories of the absurdly surreal, the surreally absurd, the existentially dreadful, and the grotesquely ludicrous. ***Stories to Make You Puke Your Pants*** is no simple pants-puking matter. These are tales to twist your mind, warp your reality, flip the script, and make you wonder why you are reading a book when you could be out playing *Risk* with your hip, young friends. You could be playing *Mario Bros.* on your Atari 7800. You could beat the game. Or you could commit to the quest for insanity and crack open these crusty pages to see what all the puke is about.

"G. Arthur Brown is a glorious minstrel making music for demon gods to dance to. His prose is taut, charming, and sinister. His mind quick, quirky, and wild. Let the moon go up at midnight and let him serenade you. Open up your heart at him: let his stories make you full." — Brian Allen Carr, author of *Opioid, Indiana* and *Motherfucking Sharks*

A Quaint New England Town
by Gregory L. Norris

When Ezra Wilson took the job as a census worker, he never imagined it would lead to a place like his latest assignment. From the moment he turns off the interstate and travels past the village limits, it becomes clear that Heritage isn't just some quaint New England town.

A sinister encounter at an automobile graveyard is only the start. In Heritage Proper, a town divided down the middle both politically and literally, Ezra is met with hostility on both sides of an imposing brick wall that separates warring factions that have maintained a fragile peace. After scaling the wall into Heritage North, Ezra discovers a beautiful young woman held prisoner in a fortified basement room and promises to help her. To do so will expose the last of the small town's dark secrets and lay bare big planetary dangers if Ezra survives his visit to a destination where even the white picket fences are not at all what they appear to be.

Russells in Time
by Kevin Shamel

Because you can never have enough Shamel! In this novella, a trio of recognizable characters find themselves travelling back in time and in the middle of a heated battle between the dinosaurs and a race of giant land-squid. Who will they side with? And will we get to see Russell Brand kicking ass in an Iron Man-esque suit? (Spoiler — yes. We totally will.)

Selleck's 'Stache is Missing!
by Charles Chadwick

Celebrated Hollywood star Tom Selleck has it all: talent, good looks, a winning personality, and a track record of television and movie hits, enjoyed by millions around the world. Until one day, while filming his latest project, an old rival attacks him and steals his mustache. Now, lost and adrift, Tom struggles with his new life. Along with a group of dedicated crew members, celebrity friends, government agents, and the robot voice of an old co-star, he has to find the strength to take on his greatest role ever: tracking down his old rival, retrieving his legacy, and saving the world.

Songs About My Father's Crotch
by Dustin Reade

A collection of bizarre tales from the author of *Bad Hotel*. There's something for everyone in this one. Yes, even you.

The Secret Sex Lives of Ghosts
by Dustin Reade

Thomas Johansson can see ghosts after a near death experience and has made a living killing them for a second time. After discovering that being possessed by a ghost causes an intense hallucinogenic effect, he goes into business with a perverted dead man named Jerry, selling possession as a street drug (street name: Ghost). But is the farmhouse he sees while possessed really a hallucination? Or is it some else?

CPSIA information can be obtained
at www.ICGtesting.com
Printed in the USA
BVHW070822100722
641727BV00014B/1965